UNEARTHING

THE

DRAGON

MARK NORELL Chair and Curator
Division of Paleontology
American Museum of Natural History

Photography and Drawings by

MICK ELLISON Principle Artist
Division of Paleontology
American Museum of Natural History

UNEARTHING THE DRAGON

THE GREAT FEATHERED DINOSAUR DISCOVERY

PI PRESS

New York   New York

2005

A PETER N NÉVRAUMONT BOOK

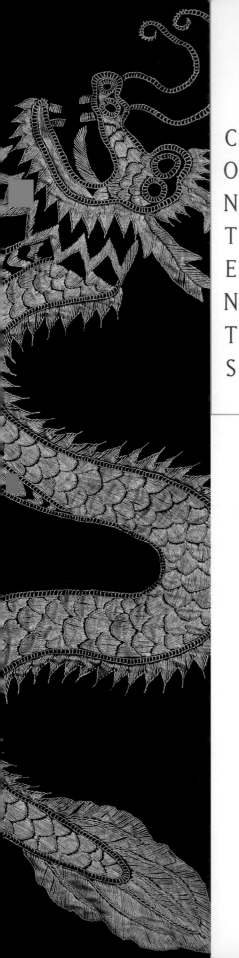

CONTENTS

PI PRESS

An Imprint of Pearson Education, Inc.
1185 Avenue of the Americas, New York, New York 10036

Pi Press offers discounts for bulk purchases. For information contact U.S.
Corporate and Government sales, 1-800-382-3419, or corpsales@pearsontech-
group.com. For sales outside the U.S.A., please contact International Sales, or
international@pearsontechgroup.com.

Printed in P.R. China.

First Printing

Library of Congress Cataloging-in-Publication Data
A CIP catalog record for this book can be obtained from the Library of Congress.

Pi Press books are listed at www.pipress.net

ISBN 0-13-186266-9

Pearson Education Ltd.
Pearson Education Australia Pty., Limited
Pearson Education Singapore, Pte. Ltd.
Pearson Education North Asia Ltd.
Pearson Education Canada, Ltd.
Pearson Educación de Mexico, S.A. de C.V.
Pearson Education — Japan
Pearson Education Malaysia, Pte. Ltd.

Produced by Névraumont Publishing Company,
New York, New York

Jacket and Book Design: Cathleen Elliott

Note to the Reader

Many Chinese names look exceedingly hard to pronounce, given the predominance of rarely used consonants, q's not followed by u's, and a whole lot more. For the names here, this should suffice. X as in Xu is pronounced sh as in shoe (with some exceptions like Yue-Xing where the Xing is pronounced sing). Q as in Qiang, Qin, etc., is pronounced like a hard ch as in cheese or chin. Z's are usually followed by h's and together are spoken like the j in the French pronunciation of Jean as in Killey or Cocteau. And j's at the beginnings of words are usually hard j's as in just or John. Thus, some of the more commonly used names here are pronounced Ji Qiang (gee chong), Gao Ke-Qin (gao kee-chin), Xu Xing, (shoe shing), Zhang Mee-Man (jeang mee-man), Meng Jin (mung gin), etc. Also, the syntax of names is reversed from Western usage, so in the case of Gao Ke-Qin, Gao is his familial name and Ke-Qin his given name.

风暴中说睡的龙

Sleeping Dragons of
the Revolution

Weak Coffee, Stale Doughnuts, and Hot Fossils

In October 1996 the Society of Vertebrate Paleontology held its annual meeting. Little did anyone know that what was going on in the corridors outside the lecture rooms would dramatically alter the shape of dinosaur paleontology and open an entirely new chapter in our understanding of the Mesozoic Era and the origin of modern types of plants and animals. It was a revolution in dinosaur science. It was a sea change in thinking that will be reflected in the imagination of any child who loves dinosaurs or anyone who goes to see a dinosaur movie or visits museums. It will change our popular culture, as well as our understanding of the history of life on Earth. What was shown in the halls was the first definitive evidence that dinosaurs with feathers had been found. Now, almost 10 years later, it seems as if there are more feathered dinosaurs than we can count and the distinction between what is a dinosaur and what is a bird is blurry. Dinosaurs everywhere have been re-evaluated, even to the extreme of considering that such familiar dinosaurs as *Tyrannosaurus rex* were probably feathered, at least for part of their lives.

Chen Pei Ji is a professor at the prestigious Nanjing Institute of Geology and Paleontology part of the Chinese Academy of Science. Soft-spoken, with a round face and thin, graying hair, Chen is a shrimp expert. More precisely, his main research uses invertebrate fossils such as conchostracans (small-shelled shrimp) to study both paleoenvironment and temporal relationships of sedimentary rock formations. But Chen, like many of my colleagues trained before specialization became rampant, is passionate about all life in deep time and is always on the lookout for great new fossils. He found one from a place called Sihetun, a tiny village in Northeastern China.

The 56th annual meeting of the Society of Vertebrate Paleontology was at the American Museum of Natural History, my institution, where I serve as a curator and chairman of the Division of Paleontology. These meetings consist of a series of scheduled presentations, enforced socializing, and weak coffee and stale doughnuts. Written abstracts of the talks are published months before, so everyone pretty much knows what the presentations will contain. Rarely is anything shocking or groundbreaking delivered from the podium. At all scientific gatherings, the real action goes on in the hall, where the hot new unpublished and unpublicized stuff is exhibited and argued over, sometimes years before it is ever published in the scientific literature.

Outside the technical sessions, Chen was showing photographs of a new fossil from the northern Chinese province of Liaoning. At first, Chen's specimen looked like just another small, well-preserved carnivorous dinosaur. But on closer examination, even in the dim light of the Hall of Northwest Coast Indians, you could see that it was an animal apparently cloaked in an aura of fuzz. The hero of J. D. Salinger's *Catcher in the Rye* and another visitor to this hall, Holden Caufield, observed, "Certain things, they should stay the way they are. You ought to be able to stick them in one of those big glass cases and just leave them alone." Chen's fossil would have depressed young Holden even more than he was, because this was what many of us had been waiting for: the smoking gun of a "feathered" carnivorous dinosaur and hard evidence for a controversial idea that dinosaurs were very different from what we thought. Things weren't going to stay the way they were.

Carnivorous dinosaurs belong to the group Theropoda, theropods for short. This important group of dinosaurs includes everything from hummingbirds to *Tyrannosaurus rex*, and is the focus of my work in China and beyond, and much of what fills these pages. Although things turned out to be far more complex (as scientists always say), Chen's photos of what would come to be known as *Sinosauropteryx* shook the world of dinosaur paleontology.

Before that meeting, my recent work had been focused primarily in Mongolia's Gobi Desert, searching for dinosaurs, especially kinds of theropod dinosaurs relevant to the question of the origin of birds. Going back and forth to Mongolia through China, and in New York, I had come to know many Chinese scientists and students. I spent a lot of time in China and loved every minute of it. I started to make professional inquiries to see some of the "feathered" specimens. At the time, I didn't anticipate being included in the research. I certainly never anticipated that in these past few years, through the discovery of so many specimens and the hard work of an international team of scientists, the idea that the dinosaurian relatives of birds were feathered, just like birds today, would be popularly accepted. This book is the story of how that idea changed the dinosaur world.

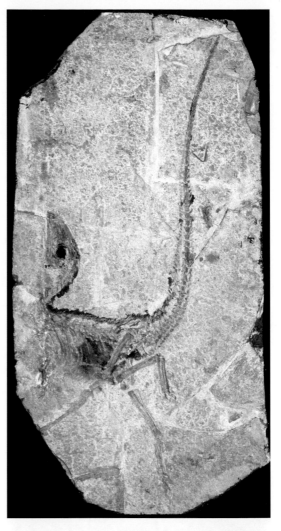

Sinosauropteryx prima, the first of the "feathered dinosaurs" to be discovered. This is a small animal about 130 cm in length. The primitive feathers are clearly visible running the length of the body above the back, head and tail.

China captured my mind and my heart. As rewarding as the science, working and traveling around China, meeting, sharing meals, and enduring lengthy drinking sessions with scientists, laborers, shopkeepers, tradespeople, local officials, farmers, and just whomever we happen to meet are deeply satisfying. Mick Ellison, who is responsible for the photographs and drawings in this book, has accompanied me through the feathered dinosaur phase of these travels, and it is through his images that we hope to succeed in bringing to life some of these experiences. During our travels, we have learned a lot, but as any Westerner who has scratched the veneer of China can attest, there is a whole lot more out there.

Many times when we go to China, it seems, something extraordinarily bad happens. Even before we land, the storm is on us. Chinese military aircraft have intercepted U.S. spy planes. On the occasion of one of our arrivals, the Chinese fighter crashed, the pilot was killed, and the wounded U.S. plane was forced down on Chinese soil, resulting in diplomatic posturing and a standoff. Arms sales to Taiwan are seen as contrary to the Shanghai communiqué, signed by Richard Nixon and Mao Zedong, which acknowledged that Taiwan is part of China. Wars in Iraq and Afghanistan have been scrutinized on the streets of small towns as cases of U.S. imperialism—as we have been vociferously reminded on those same streets. Exotic local menu items in South China led to the outbreak of SARS in the spring of 2003, the absolute stoppage of almost everything in China, and 10-day quarantines for us on our return to work in New York.

The worst event was the bombing of the Chinese Embassy in Belgrade by American laser-guided bombs in May 1999. Based on erroneous information by a now-fired CIA agent, this was a nadir in recent U.S.-China relations. I was elsewhere, but Mick faced the full fury of exploding Chinese nationalism. An e-mail that he sent me at the time read simply, "The dragon has risen and he is pissed." After a rocky time in Beijing, when disdain for the foreigner went from verbal snipping, to having hot tea dumped in his lap, to the threat of pummeling by heavily drinking locals in a small restaurant, Mick escaped to the North with Gao Ke–Qin, now a professor at Beijing University, a close friend and colleague who studied at the American Museum for six years in the late 1990s. Arriving in Beipiao, a grimy coal-mining town in Liaoning Province, things were not that much better, and in one case a very drunk local attacked with the only weapon available, his teeth, biting Mick in the face. The last thing a local official wants is for a foreign guest to be harmed in his town, so for Mick's own protection, he was put under house arrest and moved into the domicile of an official for several weeks. This gets pretty boring really fast, as could the host's favorite morning meal of steamed bread and peanuts, congee and silkworm larvae.

Mick has a sense of humor that people of all cultures seem compelled to respond to. One night, in an effort to amuse their ever more miserable guest, Mick's hosts smuggled him to a room in a nearby restaurant for some entertainment, as it was still deemed too dangerous for a real night out on the town. There, a makeshift

A Beijing policeman in Tiananmen Square on the day of the official funeral for Deng Xiao Ping. Deng spearheaded the reforms that have made China an economic power and was instrumental in leading China out of the turmoil of the Great Proletarian Cultural Revolution.

private nightclub had been set up, complete with video Karaoke, lights and beverages. Mick's hosts made an effort to attract some of the local girls walking down the street to come in, but all were disgusted by the pathetic scene of an unshaven, pink-skinned, big-nosed foreigner hanging with a few very bored locals.

The lightheartedness of such incidents pales in the light of the brutal history of China. The people of this land have suffered some of the most terrible nationwide disasters in modern history. The Opium Wars of the 19th century followed British attempts to mollify the Chinese and lure them into lucrative trading deals by making them a nation of junkies. Shortly thereafter, 30 million were killed during the Taiping Revolution led by Hong Xiuquan, who believed himself the younger brother of Jesus Christ. The 20th century began with the Boxer Rebellion, when citizens disaffected by Western social and economic adventurism in China joined a mystical sect called the *I-ho ch'uan*, or "Righteous and Harmonious Fists." They practiced traditional martial arts, which they believed made them impervious to bullets. Spurred on by the Empress dowager Cixi, who instructed the Boxers to kill all foreigners in China, the Boxers roamed rural and urban China killing foreigners and Chinese who had converted to Christianity. This elicited a brutal response from foreign economic interests, who organized an expeditionary force of Russian, American, British, German, French, and Japanese soldiers, which marched on and looted Beijing. During this invasion, much of the cultural patrimony of China was destroyed or divided as spoils.

This was quickly followed by the collapse of the Qing dynasty in 1911 and internal revolution, leading to a warlord period and then, in 1937, invasion and genocide by Japan, and more revolution. After liberation, famine and the Great Proletarian Cultural Revolution followed. The eye through which most Chinese view the world is tearfull.

Yet historically, and especially in the new millennium, China cannot be dismissed, as it historically has been, as a second- or third-class country. In paleontology, China has a rich scientific legacy, intermingled with, but separate from, our Western tradition. Vertebrate fossils have been collected and studied in China since the late 19th century. For hundreds of years before that, fossil animals were recognized as remains of living animals and studied as such; however, this work was, until recently, unknown in the West.

Right on the heels of the Great Cultural Revolution, China entered a new phase of paleontological research. This is the cover of Volume 1 of *Hua Shi*, a popular science magazine of the Institute of Vertebrate Paleontology and Paleoanthropology. Here the girls are admiring the pelvis of *Mamenchisaurus*, a giant sauropod dinosaur from western China.

In 1859, England and France forced China to open its doors to Western commercial interests. Diplomats and businessmen discovered Chinese apothecaries, where "dragon bones" (the remains of fossilized vertebrates) were commonly sold. Some of these escaped the mortar and pestle and made their way into European collections, where they were studied by early European paleontologists such as Richard Owen of the British Museum. Swedish and German paleontologists, followed by Americans, organized expeditions during chaotic times in the country and made collections from all over China. Liaoning is in the northeast. and although it was missed by these early Western paleontologists, there is evidence that it had been producing dragon bones for hundreds of years.

Two men ignited the history of paleontology in China: Yang Zhungjian and Chow Minchen. Yang was educated at Beijing University, roughly at the same time Mao worked as a librarian there, and then received his Ph.D. in Munich during the turbulent 1920s. When he returned to China, he took a position with the Geological Survey, and during his lifetime named over 200 species of fossil vertebrates and authored over 500 articles, on everything from geological aspects of fossil localities to descriptions of fossil reptiles. Born in 1897, Yang was instrumental in navigating Chinese paleontology through difficult political and social conditions of the mid-20th century, and in melding Western vertebrate paleontology with a distinctly Chinese view.

In the midst of mayhem, Yang trained a remarkable student, Chow Minchen, who would succeed him. Chow received his Ph.D. at Princeton, and for the rest of his career was associated with the Institute of Vertebrate Paleontology and Paleoanthropology (IVPP) in Beijing. Chow was a great scientist, a good politician, and an advocate for the scientists of the American Museum of Natural History. He was a dapper and worldly fellow. I fondly remember him, often sitting next to my Chinese-born wife at banquet dinners chattering away in Mandarin. From their laughter, the jokes were polite but acerbic ones, directed at me—the kind that when translated make no sense to non-Chinese speakers. Following Yang and Chow's lead, there is no question that paleontology in China today is flourishing.

China itself, after a long stasis, is growing and evolving—at a sometimes out-of-control pace. Confusion is often rampant as cultures of town and country, young and old, modern and ancient collide. China has become a superpower on the global economic and military stage. Inevitably, Westerners can only understand these social transformations through the spectacles of Western cultural traditions. Nevertheless, the life philosophies of East and West remain very different.

Chinese mythology, folklore and history are rich. It is the most common subject matter of material culture and traditional crafts thrive and are part of contemporary culture, not just objects for the tourist trade. Here a scene from an ancient legend is etched into a small egg-shaped gourd.

China has several Creation stories. One features a woman named Nu Gua, who in her 70 transformations created the entire cosmos, but without people. Soon, she became bored and lonely, so she molded figures of yellow clay in her likeness and breathed life into them. Nu Gua quickly became tired of making humans by crafting each one individually. She found an easier way. She would drag a cord through wet mud, swing it, and splatter the ground with blobs of mud, which she would animate. In this way, she populated the world with people. The special yellow ones handcrafted by Nu Gua herself were the Chinese aristocrats.

The Temple of Heaven is the best-preserved Ming monument in Beijing. The circular mound alter is part of the Temple of Heaven complex. It was on this platform emperors presented sacrifices to the gods for a good harvest on the Winter Solstice. Today, Chinese tourists jostle for position to be photographed on the spot that in legend is "The Center of the Universe."

The calligraphy of Gao Ke-Qin. Calligraphy in China is the highest of skills, and one whose practice takes a martial arts like form of repetition and repetition.

Another story is that Pan Gu created the world from Chaos. The story goes that the universe was originally formless chaos enclosed in a big black egg. After living in the egg for 18,000 years, Pan Gu separated its formless contents into the lighter Yin part of the heavens and the heavier Yang part of Earth. Pan Gu's tears formed the rivers and his breath the winds and clouds. The lice and fleas from his body are the ancestors of humankind.

How does this relate to dinosaur science? It doesn't, directly. It does begin to explain differences between the ways our Chinese colleagues and we Westerners approach science and work together. As Pan Gu's creation story suggests, in China, there is a fundamental expectation that the world is balanced, that balance is primary, and that chaos only erupts when the balance is broken. In the West, we reverse the order. Chaos is primary; the harmony of the state, or our lives, is something we create. Furthermore, as the cultural descendents of Nu Gua's handiwork, Chinese tend to feel that the rest of the world is composed of barbarians—xenophobia is not an uncommon trait in populations both East and West. Much has changed in recent years, but as in our own culture, prejudice remains. Today foreigners are often referred to as *waiguoren,* which literally means foreign people, but other terms for foreigners refer to devils and ghosts.

A divide also exists in how we do business. New Yorkers are direct, and Chinese speak in vague terms. New Yorkers do business with people we don't like and sometimes have never met. In China there is a long process of getting to know one another, usually over several meals, before doing any kind of business—including science. Behaviors commonplace in New York are often seen by the Chinese as impolite. My wife is Chinese, and we have often witnessed such cultural clashes firsthand. My mother-in-law recently admonished me for putting a used tissue in my pocket. To her, this is the equivalent of pocketing used toilet paper. But spitting on the sidewalk is fine. Both Mick and I, and our colleagues on the mainland, hope that our work and this book will contribute to bridging this cultural divide. Even though mainstream China and America have very different opinions on everything from politics to music to food to aesthetics, there is a lot more common ground.

The title of this book may seem self-explanatory. But there is something deeper. The calligraphic inscription rendition of **Unearthing the Dragon** was created in the grass style by Gao Ke-Qin. Meng Jin developed the individual chapter headings. There are a number of different ways to translate Western words into Chinese characters. Characters have specific meanings, and because the Chinese have, in at least one way, a more complex language than English, it is often difficult to choose which character is the most appropriate. Scholarship of characters and calligraphy is a living art in China, and most bookshops have extensive sections on the theory, criticism, and practice of calligraphy. In this instance, the character used for unearthing usually means to reveal, rather than to dig up, and the character for dragon is more about the essence of a dragon as a living, breathing creature than it is about a caricature incorporating fangs and flaming eyes. The importance of the discovery of knowledge and its pursuit is built into these Chinese characters.

Unearthing the Dragon is a look at how we develop new scientific ideas, and how discovering fossils of incredible organisms is the everyday work of dinosaur science. Unlike any other project I have ever been involved in, this book demanded to be written. It was not planned. I didn't go to China to seek the experiences I have had. I didn't begin my collaboration with Chinese colleagues on the dinosaur fossils of

Flagons of hot water and boxes of tea are ubiquitous parts of any Chinese home or workplace. In the cities vacuum bottles like these have largely been replaced with solid- state electric teapots.

Liaoning Province or cultivate their friendships with a book in mind. My partner in this, Mick Ellison, didn't start taking pictures of people and places in China beyond the dinosaur fossil sites and the specimens unearthed because of a plan, either. We just fell into it because we were fascinated with the fossils and the landscape, and we are addicted to the people and their culture.

Theories abound on what science is, on how progress is made, and on what it means to be a scientist. This stuff is a discussion topic for all scientists in training and the focus of many barroom discussions on the Upper West Side of New York City, with my colleagues here at the American Museum of Natural History. The general public knows little of this, instead viewing scientists as either akin to theologians in discovering the truth, or as unusual eccentric types, maybe up to no good, and usually with a colonial or imperialist attitude when working internationally. Unfortunately, this has been true far too often. My outlook, however, is more pragmatic. In *Unearthing the Dragon*, using the plants and animals found as fossils in the Liaoning rocks as a vehicle, we believe we have continued the process of international science in the close company of our Chinese colleagues and friends. During the research for this book, we have seen China and Chinese science in a period of rapid change. The book also tells the story of the prehistoric plants and animals of Northeastern China, how they relate to the organisms of today, and how our knowledge of them is a changing landscape altered by each new discovery. Ultimately, these pages identify a profound adjustment in our understanding of the dinosaur world.

That meeting at the American Museum of Natural History in 1996 was a landmark in the transformation that has occurred in this generation in our understanding of dinosaur biology and bird origins. While there have been many new discoveries in places as far-reaching as Patagonia, the Sahara, and the Gobi Desert, one discovery stands above all else: the feathered dinosaurs of Liaoning. Without recasting the history of dinosaur studies, popular culture shows us just how far we have come. In 1993 the movie *Jurassic Park* embedded in our psyche a 1990s view of dinosaurs as smart but scaly creatures. Compare that with recent reconstructions in the series *Walking with Dinosaurs in America,* where many are warm-blooded and feathered, attend their young and brood their nests. In 2004, I co-authored a series of papers with Chinese paleontologist Xu Xing. We established that the greatest icon of dinosaur studies was going to have to be redrawn. Even *T. rex*, we now believe, thanks to yet another fossil discovery at Liaoning, had protofeathers.

An exquisitely preserved dragonfly. Dragonflies have a very long history and in the distant past reached colossal size. By Jehol times, however, they were nearly identical in size and anatomy to those that patrol late summer skies today.

The Liaoning discoveries have come to be called the Jehol biota. Fossils at this locality range from about 135 to 110 million years old. Most were deposited on the floors of lakes, where they were preserved in fine-grained volcanic ash that rained down on the lakes' surfaces. Some specimens are also preserved in more coarsely grained rocks, and some seem to have been buried alive. Toxic volcanic gases may have caused mass mortality. All kinds of animals and plants have been found as fossils. Fish, leaves, and insects are the most common, but mammals, frogs, and lizards, as well as true birds, have also been discovered, along with dinosaurs. Many of these organisms are extremely important to scientists studying the origin of modern groups. Most of the mammals appear to be primitive relics of much more ancient forms. One has even been found with its last meal of baby dinosaurs preserved.

Arguably, the first (or at least some of the earliest) flowering plants are known from Liaoning, as are extremely primitive birds that retain the "reptilian" characteristics of teeth and long tails.

Beyond this, through a serendipitous confluence of factors, many of these specimens contain preserved soft body parts like scales, skin, and even feathers. While we have known for a long time that birds are the direct living descendants of dinosaurs, it was these fossils that demonstrated just how birdlike in appearance and even behavior many of these dinosaurs were. These remains have told us much about how and why feathers evolved, about the origin of flight in modern birds, about the dynamics of ecosystems in the "Age of the Reptiles" and, ultimately, about the very nature of the evolutionary process itself.

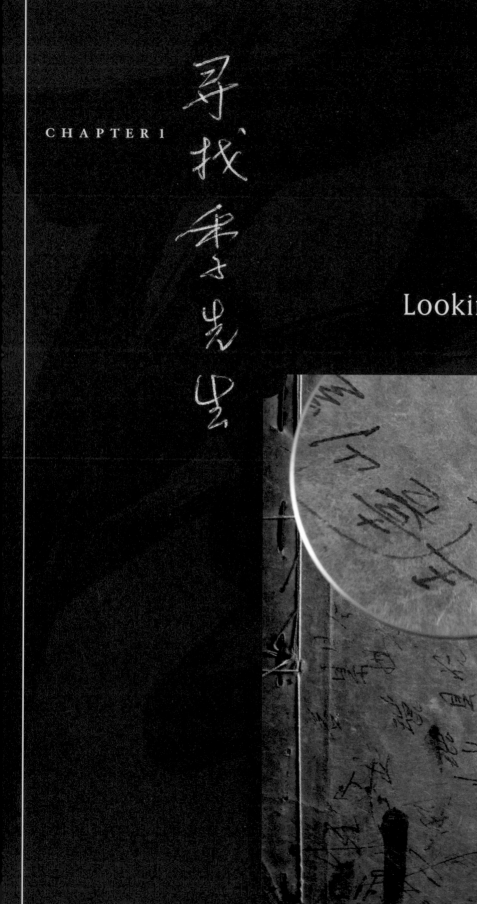

CHAPTER 1

尋，找季矢先生

Looking for Ji

"Feathered animals are bird or dinosaur?"

It had been a long flight, 26 hours from door to door. Mick was wedged uncomfortably in the center island; I was on the aisle. The passenger on Mick's right was unconscious with travel weariness (except at mealtime), her head drifting over onto Mick's shoulder in her slumber. She was young, a Beijing girl, and cute; Mick said she smelled good. He enjoyed it. Maybe I shouldn't have given him the Valium. He took one with no effect just a few hours out of Beijing, just before our stopover in Japan, and the other a mere 20 minutes later. It hit him like a freight train. Looking back, washing it down with a beer was a bad idea. The 747 landed on one wheel and shuddered to a stop on the runway, the last of Mick's peanuts spilling on his lap. I almost had to carry him off the plane. The nice-smelling girl averted her eyes from the indecorous New Yorkers. The aisle was cluttered with magazines and newspapers that had crept out over the floor—they looked like we felt. We stumbled out, wading through the flotsam and jetsam and made our way through the hatch into a cold rainy Beijing night.

The frigid, humid air had a sobering effect. We trundled our way through labyrinthine hallways schlepping our heavy cameras and computers to the packed cavernous room where one gains entry to China. The line was long, and there was a perfunctory examination of our travel documents. The "k-chuck" of the stamp in our passports signaled our entry into the Middle Kingdom.

If you are *waiguoren*, brightly clothed young men and women immediately surround you, and for a paltry fee of 6 Yuan (about 75 cents) will get a baggage trolley and escort you through customs. I tip well and up front, so our guys just crowded through everyone else yelling, and people scattered. At home, this would get you killed, but in China, there is not such thing as an orderly line. It is simply who gets there first or pushes harder. Our guys and we exploded out of customs and into the crowds. The arrival area was packed with people waiting for relatives, waving signs with people's names, and hustling cab rides. We were led through the masses by our guys. The automatic doors parted and we were at curbside.

At the curb, in the so-called queue for cabs, a couple more small bills placed our tired bodies right in the front row and quickly into a neat air-conditioned cab. The 40-minute ride into the city was spa-like and restorative. I fell into tour-guide mode, telling China neophyte Mick about all the scenery that could be seen on a cold, dark winter night. Revived, we were at our hotel before long. It is midnight here, but noon our time. Over a cold Tsingtao and spicy Szechwan noodles, we hatched our plan for the next day. We will go to the National Geological Museum of China in search of Ji—a man we had never met.

How do we know about the past? How can we determine how things like feathers or flowers evolved? We can make inferences based on the distribution of characteristics among Earth's current life forms, but the only direct evidence for the origin of these characteristics is the fossils of the most primitive animals to have feathers and the earliest plants to have flowers.

At this writing, I sit in my office at the American Museum of Natural History in New York City, looking out windows across Central Park at the lighted Manhattan skyline. I am surrounded by cabinets of fossil bones representing over a hundred years of global collecting by my predecessors. In collection rooms, out of public sight at the museum, are hundreds of thousands of fossils. This

Peering out my office window, a lofty turret perched above the west side of Manhattan's Central Park.

Zhang Mee-Man holding one of the objects of her study, the ubiquitous *Lycoptera*. This small fish is by far the most common vertebrate fossil in the Jehol deposits and is often found in mass mortality layers, perhaps the result of instantaneous death due to volcanic eruption.

collection is arguably the finest and most complete assemblage of fossil vertebrate animals in the world. It forms a reference library of objects that documents the history of life on our planet. Yet for many questions, such as when feathers originated, in what sorts of animals they first appeared, and for what purpose they evolved, the collection tells us little.

Paleontologists are always looking for new El Dorados of fossils that can shed light on such big questions. Usually the discoveries come piecemeal, one fragment at a time, or through the innovative use of some new technique or technology. So paleontologists around the world were unprepared for the onslaught of fossils that came, and continue to come, from the fossil deposits of Liaoning Province in Northeastern China. The fossils are answering important questions, such as the stages of feather evolution, but they are also changing the questions themselves: How do we define what constitutes a bird? If we use a character-based definition—such as, if it has feathers, it's a bird—then we are stuck with some very un-birdlike animals that didn't fly, had big teeth and claws on their hands and feet, and had very long, stiff tails.

During the brutal Japanese occupation of Northern China, Japanese scientists discovered extinct fishes and aquatic reptiles in Liaoning. The Japanese called the area Manchukwo, and most of the fossils studied by their scientists were lost in the chaos of Japan's defeat in World War II. Since the establishment of the People's Republic of China in 1949, it has been primarily Chinese scientists who have researched the material. Notably, fossil fish from the area have been studied by Chinese academician Zhang Mee-Mann of Beijing's Institute of Vertebrate Paleontology and Paleoanthropology. Mee-Mann is the doyenne of Chinese paleontology and a frequent visitor to the American Museum of Natural History. Although she would never take credit for it, she is directly responsible for the training and internationalism of the current crop of excellent young Chinese paleontologists, which has made China a world power in paleontological research. Mee-Man is self-effacing and refers to herself as "an old retired lady who studies fish." But this has as much truth to it as Chinese women saying that Western women are much more beautiful.

In the early 1990s, new specimens from Liaoning began to emerge. Unlike the earlier discoveries, some of these were fossils of terrestrial—land-living—animals. Students from Beijing, such as fossil bird expert Zhou Zhonghe, had begun collecting in the Liaoning area, finding one enticing specimen after another. First, a primitive toothed bird was found, then a slightly more advanced one, and then reports of dinosaurs appeared. Rumors began to circulate about even stranger animals.

At this point it is important to point out that because birds are the descendants of dinosaurs, they are dinosaurs (just as humans are mammals). Throughout the book, I will use the term non-avian dinosaur to refer to those dinosaurs that are not birds. This is analogous to the Aristotelian logic of dividing humankind into Greeks and barbarians. While all were human, the barbarians were composed of myriad races and tribes, and the Greeks (and in our analogy here, the birds) are only a subset of the larger group. One of the major groups of dinosaurs is called theropods, as mentioned earlier. And one of the subgroups of theropods is called coelurosaurian dinosaurs. This subgroup includes familiar dinosaurs like *T. rex* and *Velociraptor,* as well as birds.

As the first technical scientific papers appeared in the mid 1990s, it was apparent that the ground in dinosaur science was shaking, and my interest immediately turned toward the Liaoning fossils. They formed a new chapter of the very complex story of bird origins, a story that I have been diligently pursuing for the past 15 years, and one that is intimately tied to my other ongoing work on close bird relatives, the theropod dinosaurs of Mongolia.

In March 1997, Mick Ellison and I journeyed to China to examine new dinosaur specimens from Liaoning. It was Mick's first trip to China. He had heard my stories about it for years, so he expected a lot. He and I had worked together since 1990, when I hired him as my artist and assistant at the museum. Mick is not a scientist; he went to art school. He is a careful observer, and one of those traveling companions you would go anywhere with. We look at the world in many of the same ways, and crucially, given how much of it we have done together, are simpatico when it comes to traveling. His photography and observations have made my research much of what it is.

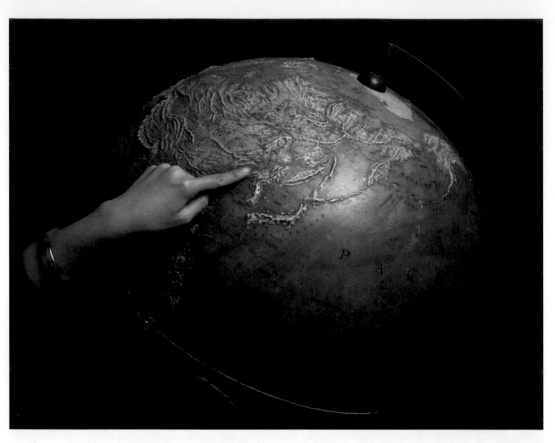

Liaoning is a coastal province in North Eastern China.

This trip began our work on the fossil animals from Northeastern China, animals that lived over 100 million years ago. The Jehol Biota is named after the city that was the former seat of the Qing emperors' summer palace. The way to think of it is as a community of plants and animals, the remains of which are preserved in sedimentary rock layers.

The morning after our arrival in Beijing, we got up early. I took Mick on a quick walking tour of the Sanlihe neighborhood in the northwest. Hung over, jetlagged, and famished, Mick grazed his way down the narrow streets choked with ephemeral breakfast stalls, sampling everything from *youtiao* (a doughnut-like fried bread), boiled peanuts, *jian bing* (a crepe with scallions and chilies cooked on a hotplate affixed to the back of a bicycle), and thousand-year-old eggs, to noodles and *jaozi* (steamed pork dumplings). Mick was taking it all in, one bite, cigarette, and photograph at a time. All I had to do was find the big stupa.

The fossil specimens we had heard about were under the control of Ji Qiang, then at Beijing's National Geological Museum of China. We had never met Ji Qiang. Tentative arrangements were made through a third party, Gao Ke-Qin, who was still working in my research group at the American Museum. The National Geological Museum is in western Beijing, in one of the more traditional and less developed areas of the city. It's fitting that the closest landmark to the museum is a large stupa constructed during the Ming Dynasty. The imposing white structure is located within the gardens of what the locals refer to as the "Bird Temple." Stupas are Buddhist reliquaries, big solid objects that house bones, hair, or teeth of the highly enlightened. We steered toward the stupa, as we discussed for the hundredth time what kinds of relics bearing on bird origins we might find at the museum. Despite clear, emphatic directions provided by Gao Ke-Qin and our hotel concierge, neither the museum nor the stupa was where we thought it was supposed to be. Asking for directions was beyond our language skills. We wandered around, snacking, and eventually we stumbled on the place, more or less by accident. Its entrance is down a narrow hutong, or alleyway, nestled among courtyard residences, small restaurants, peddlers, bicycle repairmen, and shops.

Two specimens of *Confuciusornis* preserved on the same slab. *Confuciusornis* is a primitive bird that bridges the morphological gap between *Archaeopteryx* and more modern avians. *Confuciusornis* is the most common avian in the Jehol deposits.

The lively world of China street food. Here a *youtiao* dealer peddles his product.

A museum assistant met us, dressed in an old blue Mao jacket and drinking tea from a jar. After about 10 minutes of our trying different pronunciations of Ji Qiang's name, he finally got the point. With the assistant endlessly cackling "Ji Qiang, Ji Qiang" and laughing at our fractured *Beijing hua* (the capital dialect), we were led through dark corridors to a small subterranean office, a damp collection space where we found Ji Qiang sitting at a low table wearing reading glasses and making notes with a pencil on paper.

"Hello," he said, surprised. As it was nearing the noon mealtime, he motioned for us to come with him for lunch. It was still winter in Beijing, and the elm trees looked like skeletons. West Beijing is still an older part of the city, and in 1997 it was very much grayer than today, so much so that the old Qing buildings and the newer post-liberation ones seemed to meld with the cold February skies. We left the Geological Museum, collecting a whole entourage of Ji's co-workers as we passed dark office after dark office. We swept past the gigantic quartz crystals from central China, monolithic limestones from the karsts of Guangxi, and the petrified logs from Xinjiang that guarded the main entrance and walked down the street to a small restaurant. Ji Qiang was obviously a regular, as the pair of *ni hao* girls (women in tight red *chipaos* who stand by the doors of restaurants to greet customers) swept open the doors and giggled as the hostess took us to a large, round table in the back. No need for a menu. Ji just rattled off the names of a few dishes as the first round of tea was poured and cigarettes lit. Everything in this restaurant—carpets, walls, menus, uniforms, and lipstick—was all the same incandescent red.

All of us were uncomfortable and awkward. In accordance with Chinese manners, we sipped our tea and began to make small talk in slow drawn-out sentences of short words. It didn't take long to find out that Ji Qiang speaks English and German quite fluently. We didn't want to appear too presumptuous about whether he would permit us to see and photograph his collection. We knew that being pushy would get us nowhere. The specimens were, after all, unpublished, and Ji had every right to be suspicious. Having such scientific intellectual property ripped off doesn't just happen to Indiana Jones. It is, unfortunately, commonplace in the real world.

During our lunch we learned that Ji (who had been largely unknown to me) had received his Ph.D. in Germany and had worked with mutual friends in the world of invertebrate paleontology. He told us that his research into his new collection of Liaoning fossils was his first foray into the realm of backboned critters. We also discovered that Ji had already described the specimen of *Sinosauropteryx* in a Chinese publication in 1996 that went largely unrecognized in the West. This was two years before Chen Pei- Ji, Dong Zhi-Ming, and Phil Currie gave the specimen additional attention in *Nature*.

The food was served in the usual Beijing style, in several multipart courses. Pausing between dishes, Ji sat back and lit a cigarette. Talking as smoke spilled out of his mouth, he asked me directly, "Feathered animals are bird or dinosaur?" The following animated three-hour discussion about how groups of organisms are recognized and defined put all of our social and scientific skills to the test. It is fortunate that Mick is capable of drinking enough beer to drown an Australian and has put enough money into the coffers of Southern states through tobacco taxes to make him an honorary Confederate. I guess we passed the test. Later that afternoon, a new world was revealed to us, as specimen after mind-blowing specimen was produced for our inspection. And so, our work and friendship with Ji Qiang began.

On this first day at the museum Ji showed us—and allowed us to photograph—the crown jewels of his growing collection, including specimens of some very feathered dinosaurs. These were the first fossils to be found of animals more primitive than

In some parts of Beijing, a quiet traditional life is still to be found down narrow hutong alleyways.

An ebullient Ji Qiang in the main quarry at Sihetun. Ji has been our close collaborator and good friend on a number of Jehol projects.

birds that show definite evidence of feathers. I leaned closer over the satin and velvet boxes that held the fossils. Although Ji had a good microscope, it was attached to a homemade stand, and the illumination was a decaying fluorescent desk lamp. I feared that a wrong move would either electrocute me or send the microscope crashing down on one of the most important new dinosaur specimens in the world. But I was so excited by what I saw: There under the flickering green light were feathers structurally identical to those of modern birds. I was looking at the origin of one of the most beautiful objects the fossil record has preserved, something that enabled warm-blooded creatures to dart across the sky and migrate over oceans and continents. This specimen conclusively showed that modern feathers didn't evolve for flight at all; they were present in animals much more primitive than flighted birds, animals that were usually pictured as more like big lizards than strutting peacocks.

Over the next week, huddling for warmth over our halogen lamps in the National Geological Museum's refrigerator-like basement, we studied and photographed the specimens. We had the initial problem of converting this space into a multipurpose photo studio and research room, where Ji Qiang and I would study specimens and type our notes while Mick photographed others. Our first challenge was to block a wall-sized window that flooded the room with low-angle late-winter light.

Eight years ago, back in the dark ages before high-resolution digital cameras, going on the road to photograph was like a 19th century Grand Tour. Bags of cameras, tripods, lights and their stands, extension cords, clamps, rope, reflectors, film, draping etc., all had to be compressed into enough luggage to avoid overweight and excess baggage charges, a big expense on our small research budget. We would go to the airport with duct tape to make two camera bags into one if the person behind the counter questioned our allotted bag number. But this stuff was important, and we were able to use the draping to completely block out the light and turn Ji's collection room into a photo studio.

The basement sessions resulted in a *Nature* cover photograph by Mick of the fossil feathers with a reconstructed *Caudipteryx* superimposed on it (June 28, 1998). The article by Ji Qiang, his student Ji Shu-An, our Canadian colleague Philip Currie, and me was simply titled "A Dinosaur with Feathers." Berkeley paleontologist Kevin Padian wrote an accompanying critique of our paper in the "News and Views" section of the magazine. In it, he observed, "The work of Ji et al. should lay to rest any remaining doubts that birds evolved from small coelurosaurian dinosaurs."

Since the spring of 1997, we have run up considerable frequent-flyer miles on what seems like countless trans-Pacific trips. The work is still in progress. The old Geological Museum and the IVPP have been completely renovated into modern research facilities. We have come to know Beijing better than any North American city other than New York. Ji Qiang and his crew now occupy several rooms of the Chinese Academy of Geological Sciences. No longer are meals shared at the red restaurant. At a new place near the CAGS, we linger over the remnants of our lunch, and Ji and I reflect on just how much has changed in dinosaur science over the last decade. Then we go bowling.

Beijing is a delightful place. Here we relax with friends in the fashionable part of the city, Ho Hai. This was the hutong ghetto, much of it built in the 16th century, that was slated for demolition before a plethora of clubs and restaurants occupied the scene in the late 1990's.

CHAPTER 2

Home of the Dragon

Jin Warriors, Panda Smoke, and the Millions of Years of Jehol

On the grounds of our feathered dinosaurs, nearly 400 years ago, the Jurchen tribes organized themselves into an independent state called Jin, in what is now Northeastern China. In 1644, armies of tall Jin warriors stormed out of the Manchurian plains and captured Beijing, to end the faltering Ming dynasty. For the first time since the Mongol princes established the Yuan dynasty in the 13th century, China was ruled by foreigners, the Manchus. The Manchu leaders founded the Qing dynasty, the final imperial dynasty in Chinese history.

Unlike the Mings, the Qing emperors were not Han Chinese, the largest ethnic group in China; they were Manchurians. Even late in Qing history, the aristocracy still clung tight to its Manchurian roots, retreating each summer to what is now Liaoning Province and to a second forbidden city, Shenyang, today's provincial capital. Now, nearly 100 years after the last Qing emperor, this area, apart from being the most important dinosaur fossil region in the world, is the packed, polluted center of China's rust belt, rich in agriculture, mining, and discontent.

Many of the industrial cities in Liaoning are dusty, gray polluted places.

Government offices in the region are notorious for corruption. In 2001, the vice mayor and the public property chief were executed for taking bribes and gambling away public money in Macau casinos. The mayor was also involved, and when his house was raided, investigators recovered over a hundred Rolexes, $6 million worth of gold bars, and stacks of fake antiques hidden in the walls and beneath floorboards. His death sentence (along with that of several others) was suspended to life imprisonment because he admitted his crimes and showed remorse, and he died in prison two years later. Throughout all this, Manchu nationalism lives, and violent strikes and protests swelled by bourgeoning unemployment and poverty are ongoing events in the capital.

Shenyang can be a fun city too. On the second night of our second visit, we grabbed a cab to go hear some live music at a local hipster bar. The car had the usual accoutrements of a Chinese taxi—at least one outside the major Westernized metropolises of Beijing and Shanghai: the liter-sized mason jar of green tea, stained by thousands of recharges; a rearview mirror used as a display hook for old rubber bands, pieces of string, and a red amulet of Guan Yin, the goddess of mercy; a greasy foam container with remnants of the driver's last meal, all under the serene gaze of a picture of Chairman Mao glued to the dashboard.

Such pictures are not necessarily overt political statements, as many Chinese hold Sun Yat-Sen, the revolutionary who worked to overthrow the Qing dynasty, in as high regard as the man who is commonly perceived in the West as the architect of modern China. Instead, pictures of Mao are talisman and protector, sort of like a Virgin Mary statue adorning the dashboard of your Italian neighbor's Lincoln. Just as Catholics flock to Lourdes, millions of Chinese a year visit Mao's birthplace of Shaoshan.

In fragmented English, our driver engaged us with familiar questions. "Is your first trip to China?" "How you like?" "How much money do you make?" (That's not a rude question here.) "How old are you?" "Do you like Chinese food?" Our destination was reached and the fare paid. As we exited the cab, the driver, instead of saying "*zai jian*" (goodbye), gave us a little local flavor. With a fervor that surprised us, he proclaimed, "One day we will use all the planes and bombs built in Liaoning on Beijing. Hundreds of years ago, the Great Wall was built to keep us out. We will return."

Before there were Qings and Mings, before there were Manchus and Hans. Even before there were humans or the familiar plants and animals that exist today, this area of China was ecologically very different from the way it is today. It was an area covered by forests, riddled with lakes, and populated with a rich variety of animal and plant life. Far off to the west were large volcanoes, intermittently raining ash from the sky that was deposited as fine-grained particles on the bottoms of ponds and lakes. Local volcanoes oozed lava. This was Northeastern China during the Early Cretaceous period, between 110 and 130 million years ago.

This interval spans the boundary between two important geologic time periods, the Jurassic and the Cretaceous. Fossils from around the world tell us that this 20 million-year span includes the earliest members of some of the groups making up today's biota, such as modern mammals, birds, snakes, and flowering plants. So a group of fossils of organisms that all died relatively close together in the same place and time provide evidence about how different groups of organisms are related to one another; about the biological transitions between different groups—evolution's elusive "missing links"; and for examining one of evolution's most enduring and intractable problems: How and under what circumstances do novel, complex features arise? Say, for example, feathers.

Aside from his political presence, Mao Zedong's image is also both a talisman and a pop image. Here is one of several curio stores in Beijing that specialize in relics of the chairman and one of the treasures garnered there.

The fossil localities lie over a broad area, and it has not been possible for us to visit them all. In fact the fossil-bearing rocks extend as a broad sweep over Northeastern Asia. In the east they begin in North Korea, where rumors of important fossils abound. The beds traverse into Liaoning and continue for hundreds of miles into the Inner Mongolian Autonomous region. I have been able to visit all the most important sites, which has given me a great appreciation for how hard the local villagers work to collect the fossils, as well as how dangerous it is for driver, passenger, and pedestrian to navigate rural Chinese roads.

The fossil localities are named after the adjacent towns or villages. The hottest area forms a rough polygon defined by the cities of Shenyang, Jinzhou, Chaoyang, and Beipiao. Each of the small farming villages is different, in the way that much of the rural United States was 50 years ago, before Americans became nomadic, moving around the country to new homes and jobs. This process is starting to affect China similarly. People are on the move, as they are lured to cities from the countryside in the hope of finding jobs. It is thought that nearly 100 million people have left farming jobs that have become redundant or disappeared due to urban growth.

Outside the major cities and towns, everything is agricultural, with sorghum and other grasses being the seasonal crops. Although things in China have changed in the past several years, agriculture in most rural areas is still barely mechanized. Many farmers till fields behind oxen, crops are harvested using long hand scythes, and the grasses are threshed by throwing the stalks on the road so cars run over them. Between passing cars (and sometimes right in front of them), people dash out and sweep the grasses to the roadside, where they use huge wooden rakes and throw them into the air, blowing the chaff off in allergenic clouds. The harvested sorghum is loaded on top-heavy wagons pulled by mules, which become yet another road hazard.

There are few easy ways for foreigners to travel in China off the major train routes. The most common is to hire a driver with a car, paying his day rate, hotel, food, and fuel costs. That's where we were when we arrived in Shenyang in the early spring of 2001, on a visit to see some new fossil localities. Most drivers we have hired in China have been safe and careful and their vehicles spotless, if not mechanically steady. A few of these guys have been psychotics. It was our lucky day. Mr. Shen was our driver, and we almost lost our lives. Sanity dictates that when you're traveling at high speed on a rough rural

Guan Yin is the goddess of compassion. She is the female embodiment of the boddhivista Avalokiteshvara and hears the pleas of women and those in danger from water, demons, fire, or sword.

Although China has changed much in the last several years, life is still very simple for these elderly men.

Small spirit doors are built into grave mounds. Such doors allow ghosts to come and go as they please, often moving in the realm of the living, their presence is often felt.

road and you see people crossing, enter a village, or encounter livestock or oxen-drawn wagons, you should slow down. Not this guy. Any movement indicating vehicular or pedestrian traffic ahead had the effect of adding pounds to his already leaden foot.

You would think that passing the scene of a middle-aged woman who had been hit by a truck and was spread over several feet of road would give Mr. Shen occasion for contemplating how dangerous it was to hurtle down a narrow roadway, choked with all imaginable sorts of traffic, in a poorly maintained microvan. Either our driver was too stupid, too drunk, or just too crazy for this to be a lesson. One very bad passing job resulted in the side rearview mirror being stripped from our microvan, probably embedded bug-like in a truck radiator. When we arrived in the village of Sihetun, a Mecca for dinosaur paleontologists, I would have given anything to hail a New York City cab home.

Sihetun quickly had a calming affect. It is a beautiful village nestled on a seasonally dry watercourse forming the main road through town. Entering the town, we passed a small cemetery composed of large earthen beehive-shaped cones. Each has a small door at its base, a ghost door to allow the spirits to come and go. In Chinese tradition, ghosts of ancestors exist very close to the living. Offerings of ghost money are still burned on important occasions in traditional households.

Compared with those in nearby villages, the white houses of Sihetun are well-kept and large, shaded by huge cottonwood trees. A big cloud of dust, with chickens scurrying in all directions, announced our arrival, but people hardly noticed. Sihetun is used to visitors. The fossil site right outside the village is the most famous in all of Liaoning and is where farmer Li Yinfang excavated the first specimens of feathered dinosaurs in 1994. He has become a local celebrity. He lives in a white, stucco house near the entrance to the village. Proceeds from the fossils and income from those who have come to study them account for a number of snappy houses.

We last visited Li and his family in the spring of 2004. Our old friend treated us to the same sort of simple lunch of chicken, mild peppers, hardboiled tea eggs, peanuts, steamed bread, and an assortment of early vegetables from the community hothouses that is typical in this area of China. Li is the same, although his daughter is now a young woman who spends most of her time away at school. Conversation was more about the lack of early spring rains than about the fossils from the nearby hill.

To dinosaur paleontologists, this place is one of the wonders of the world, but although fossils have made some people in the village relatively well off, Sihetun is still a mixture of naiveté, promise and problems. It reflects modern and changing China. A new school has been constructed, and many of the houses in the evening are bright with electrification. A road is being built that will carry visitors right to the fossil quarry. But there are very few men in the village, as most have departed to the cities trying to find higher paying jobs as laborers, thus joining the burgeoning ranks of China's migrant laborers. Traditionally, this was unheard of; many families can trace their familial lineages for generations in the same local villages.

From the perspective of Mick and me, coming from a culture where people drive lawn mowers to cover a postage-stamp-sized front yard, we observe with mixed emotions 60-year-old women packing newly planted fields by dragging large stone cylinders behind them. I don't know if our lifestyle is much better; it is

Li Yinfang sitting in the quarry at Sihetun.

ecologically less sustainable. On our most recent walk down the narrow valley from the fossil quarry to town, we stopped for tea with some local matriarchs. Aside from admiring my silver bracelets (which were collected from loved ones years ago and our now part of me), their big question was how Mick and I could leave New York and come to Sihetun this time of year. Didn't we have crops to be put in? Such is China.

The excavation itself is a vast pit, defined by two perpendicular walls. The walls are marked with Chinese characters, an attempt at bookkeeping, showing the exact sedimentary levels where different specimens have been found and geologic samples were taken. Wide, dark layers represent lava that was deposited between the sediments, laid down by the volcanic ash that buried animals and plants in lakes and ponds millions of years ago. Mining for these fossils, both clandestine and sanctioned, still goes on here. To discourage the former, a concrete tower, housing armed but usually absent guards, looms on the hill above. When we visited the site in 2004, we were amazed to see the excavation had increased in size over three times, and that now a large metal and glass (the favored building materials of the new China) structure had been built over a section of the quarry. This is an attempt to create an in situ exhibit akin to Dinosaur National

The new building in the expanded quarry in Sihetun. The building sits on the quarry floor and abuts against the quarry wall where many specimens are preserved in place.

Monument. It is a big blue building cast against a gray dusty landscape, with nothing for miles around—it will be interesting to see if it becomes an attraction.

During our 2001 visit we took a quick tour of the site, and we moved on. We had been invited to the Beipiao Museum about 50 kilometers away. It is here that many of

Technicians working on a slab containing multiple specimens of the primitive avian *Confuciusornis* in the renovated preparation laboratory in the Beipiao Museum.

the important fossils from Sihetun and other sites are housed. I opted for a seat in the back of Mr. Shen's microvan on the theory that, at the speed which I knew we would travel, the chance of a rear-end collision was nil. The museum (now replaced by a much more modern structure) was tiny and located about 100 meters up from a trash-filled canyon carved out by the efflux of an open sewer pipe running down the hill. A strange site now is the presence of new surreal Bavarian style cottages sitting on top of the former dump.

Even in initial viewing, many of the specimens on the Beipiao Museum were among the most extraordinary that I had seen. Fish with skin patterns still preserved, several species of fossil birds, and lots of small dinosaurs, with hints of feathers. After our harrowing drive, Gao, Mick, Ji, and the rest of our crew were exhausted and in no frame of mind to spend hours looking at specimens. Back in those days, the superb specimens were not particularly well cared for, and we had to pick through piles of tantalizing fossils, stacked like old books, to continue the search. Each time we dusted off a slab of rock in the dim light, parts of fragile bones and feathers scattered in the wind.

The museum's only technician was removing matrix, the rock that surrounds the fossil bones, from a specimen of a small dinosaur, using a large rusted nail wrapped in tape. Even more surprising was the sight of several specimens of fish, reptiles, and plants for sale in the "gift shop," which was just a glass counter with fossils stacked among cans of beer, water, and ginseng soda. All this has changed now, and the quality of preparation and the organization of the gift shop have been significantly improved.

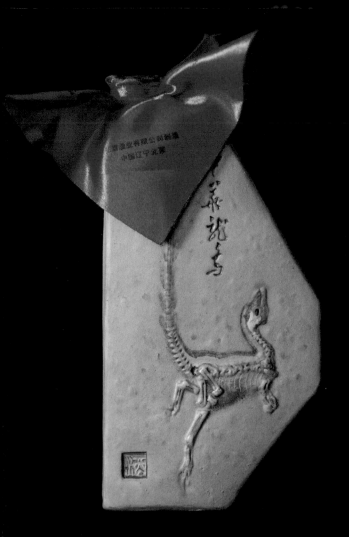

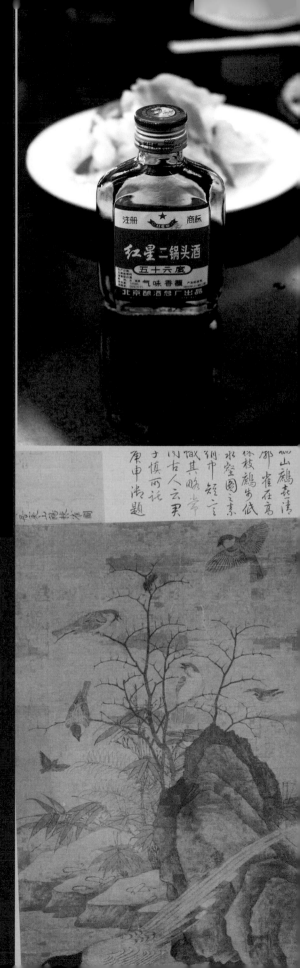

Most banquets are punctuated by toasts of bai-jiu. There are many local brands, even one in an inscribed commemorative bottle with *Sinosauropteryx* in bas relief. Here Lao Wu (Mr. Number 5) supervises the pouring of a cup.

That night, members of the museum, the Liaoning fossil protection organization, and other local officials took us to dinner. Years of experience made me painfully aware that the way Mick and I performed at dinner would go far in determining just how much access we would have to the specimens at the Beipiao museum. As soon as we arrived at the restaurant, I knew we were in trouble. We were separated immediately from our younger friends, a film crew sent to do a documentary for CCTV on our work with Ji Qiang, and hustled into the back room. The room was very yellow, as I recall, illuminated by a single florescent light whose pulse was visible as black heartbeats running the length of its dusty glass body.

Having spent time in China, and having been exposed at home to its culture through my wife and her family, I was extremely confident that I could smoothly handle a hardcore Chinese dinner session, without seeming too rude, dim, or just plain American—but this was as good as it gets. No matter where you are, from Chengdu to Canal Street, there is always discussion, confusion, and questions about the menu in a Chinese restaurant. I don't know why most places have a menu, because everyone wants to order a dish that's not on it. This night, there was no banter. Our hosts circled into the room and slapped sunglasses, cell phones, cigarettes, and lighters on the table.

We were in T-shirts and jeans; our hosts in cheap dark suits and ties. Mick and I were tired and overheated; everyone else was clean, cool, and relaxed. Adrenalin metabolites filled our veins from the hundred or so near-death experiences at the hand of Mr. Shen we had had on the road that day. All I knew about China told me this was bad. This had nothing to do with eating. This was about "who was better" —a flat-out challenge that would put our years of experience to the test.

Mick was on my right; Ji Qiang on my left. Already, yellow smoke from Panda cigarettes created a fog bank in the room. Pandas were the favorite brand of the former leader of China, Deng Xiao Ping, so you could say there was a sense of governmental power being present. Across from me sat Owl Man, so named by Mick because his large eyes were separated by so tiny a distance on his pock-marked face.

The Northeasterners are big, and at least to my eyes, many in the home team seemed to bulge out of their dark suits. By the door stood the biggest of all, a Chinese version of Mr. T. It was nighttime already, but this guy, who filled up the door, still sported sunglasses. He couldn't possibly see much, because both lenses still had the factory stickers reading, "UV professional" and "perfekt for sun and dirt."

More cigarettes were smoked, introductions made, and pleasantries exchanged. Business cards were passed around and inspected. There is no sound in Chinese for a double "l" at the end of a word, so my name becomes nor-e-er, with a lot of laughter. The first dishes came: steamed buns, boiled peanuts, and cold beef tendon. Then we got down to the real business. First, two big clay amphorae were plopped on the table. The twisted and tied ropes serving as a carrying basket were excitedly torn off, revealing cocoon-shaped cylinders, like a pair of naturalistic radioactive containment vessels. There was no lid or cap; instead, a screwdriver was used to knock the tops off the bottles with all the dexterity of a Manhattan sommelier. This was baijiu, and these guys were professionals.

Those Occidentals who have tasted baijiu always remember it. It is not a taste that grows on you. The first time I tried it, it wasn't so bad. Now I can hardly tolerate its scent from across a room. But different strokes for different folks. My colleagues and almost every male we run into in China love the stuff. It is to China what tequila is to Mexico, or rot-gut whiskey to the old West. As in the days of prohibition moonshine, every hamlet has its own brand. Distilled from sorghum, it is flavored regionally with herbs and spices. There are low- and high-proof and medicinal varieties dispensed from countertop decanters. The clear-to-yellow liquid preserves snakes, scorpions, lizards, bats, and an herbarium's worth of plant samples. It is supposedly good for what ails you, though my experience with it would indicate the opposite. The sorghum-based agriculture of Liaoning makes the province home to several famous baijiu brands. The locals are so proud of their product that they have even named one brand "Feathered bird-dinosaur baijiu." Packaged in a commemorative case, the bottle features a bas-relief of *Sinosauropteryx*.

During baijiu drinking bouts, the toasts come fast and hard. Someone is stared down. A hard bang of the teacup, and vessels are raised, adulations exclaimed, then bang *ganbei,* throw it back. *Xie xie* (thank you, thank you). Sit down, hold it down, regroup, and see who the next target is. Things get really bad when several people gang up on a few, as when there are just two round-eyes at a table of nine local hosts. What can you do about this? Not much, as it is not mean-spirited, and to gain the trust of guys like this, you have to play.

Yet there are a few tricks of the trade. First, you need to develop a pretty strong anti-gag reflex to avoid the embarrassment of what we call a Roman event. Remember, this stuff tastes bad. Second, late in the evening, you can get away with a couple of ploys. For example, if there is enough smoke in the room to obscure visibility, just miss your mouth and toss a cup full of baijiu over your shoulder. Or better yet, if your neighbor is not paying attention, quickly dump the contents of your cup into his. Unfortunately, no matter what your strategy, you will ingest some. If you have had too much, and any is too much, the best thing to do is get to your hotel room, throw up and then drink a lot of water. Hours later, when your eyelids finally part, you will hurt, but you will live.

I was thinking about a Jimi death (as in Hendrix and Morrison drowning on their own vomit) when the door opened. The Chinese Mr. T stood aside to let in someone, who, through the purple haze, looked almost as big as our security man. It was the owner of the restaurant. She wasn't fat, just huge. She made her way to the table, elbowed Owl Man aside, and sat down. It wasn't Mick and I who were about to be slaughtered; it was our colleagues and the locals who were in her sights. My spirits lifted. Instead of a mere teacup, her weapon of choice was a tall juice glass. She started late but caught up fast. Owl Man and his brothers were not going down easy, and no way were they going to be out-drunk by a woman, especially in front of foreign guests. Over the next two hours, though, they all fell. The outcome of the contest was conclusive, and the owner of the restaurant was unvanquished. Our party straggled out of the restaurant; the sensible in search of water.

Mick and I felt good enough to attempt rehydration with a warm local beer or two, over which we discussed our strategy for the rest of the trip, knowing that the later we stayed up, the faster we would metabolize the poison. We engaged in a favorite pastime of reading fractured English, literally translated from elegant Chinese on whatever was available. Curiously, it happened to be a pamphlet advertising a local Baijiu brand:

> Profound Chinese drinking culture just like Chinese History & civilization spreading far and wide. Heavily drink the enjoyment from returning to the nature and personal experience from returning to original purity and simplicity. Profound drinking culture evokes romantic imagination.

Sums up the evening well, we agreed.

During the 2001 trip, we toured for days around Liaoning, and had more run-ins with Owl Man and his crew. We also saw a lot of remarkable fossils. Besides a swollen liver, this trip to Liaoning gave me an appreciation for how different the fossil sites are from one another, how complex the region is geologically, and how beautifully preserved many of the fossils are.

The fossils of Liaoning are the result of a rare series of events. Fossils usually form when an animal dies and is buried by sediment, most often carried through the action of water. The time between death and burial is crucial for fossil quality. The shorter it is, the less time for other animals to feast on the carcass, for the elements of the skeleton to be disassociated, or for the action of aerobic microbes to reduce the skeleton to its constituent elements.

The Liaoning fossils are so spectacular because the animal or plant was buried immediately after death, or death was actually caused by the burial.

The exquisite preservation of some fossils even extends to fine details of soft tissue. Here the delicate scales on the hind limb and tail of the champsosaur *Monjurosuchus* are visible adjacent to the skeleton.

That is not to say these fossils are easy to study. Some of the most spectacular, those in which feathers and scales are preserved, are those that were smashed flat, so that what is revealed is only sections through the bones. Imagine a pigeon. Place the bird in a very thick book. Now, slam the book closed and drive a truck over it. Leave it to dry for a month, and then pry the pages open where you stopped reading. The result will be very thin pieces of pigeon on facing pages. Such are many of the Liaoning specimens. But because the bones are often visible only in cross-section, it is very difficult to figure out what the outside surfaces or arrangements of the bones were like.

The cases of soft tissue preservation are among the most interesting. Here, hair, feathers, scales, internal organs, and muscles are present as impressions or films on the slabs, intermixed with fossilized hard parts. This sort of preservation also accounts for the great diversity of soft-bodied invertebrates and plants found in these deposits. Unlike the hard parts, however, the preserved soft parts are not the products of mineralization. Mineralization is a replacement process that takes place on a molecular level, with such a degree of efficacy that even minute microstructural details are preserved. At an atomic level the calcium phosphate in bone and the cellulose in wood are replaced, often with silica or calcite. The result is an exact copy of the original bone or branch. The cavities for cells that form bone and annual growth rings in bones and fossil wood can be seen almost as well in these millions-of-years-old animals and plants as in freshly killed ones today. The preservation of soft parts like skin and feathers is somewhat different. In most instances, what we are seeing as soft parts in the Jehol specimens are not mineralized soft parts, but byproducts of bacterial decomposition. Specific anaerobic bacteria deposited their metabolic products in the exact form of organs, feathers, and hair. Because this is not as exact a process as mineralization, many times feather or hair coverings, or muscles, look like halos around specimens, and the boundaries of internal organs are ill-defined.

The fossils of Liaoning can be grouped into faunas, which are specific mixtures of animals and plants found together. Several faunas have been identified: the *Confuciusornis* avifauna, the *Cathayornis* avifauna, and the *Psittacosaurus* dinosaur fauna. These individual animals will be discussed in detail later. Here, it is enough to know that each refers to a particular ecological community, or suite of animal and plant species that lived in the same place at the same time. Manhattan is populated by the human-rat-pigeon-ailanthus-tree fauna, while in Antarctica there is the penguin-leopard seal fauna.

In the best-case scenario, faunas can be ordered temporally by determining the degree of primitivism of the various animals that comprise them. Primitivism means that an animal has retained significant characteristics of its earliest ancestors. It is important to note that primitivism does not imply inferiority and is very subjective. Some of the most primitive animals living today, such as cockroaches, are extremely successful. Other animals, such as the aye-ayes of Madagascar and Australia's kangaroos, are highly specialized, even though they belong to so-called primitive mammalian groups.

Ordering faunas temporally, based on their degree of primitivism, has its problems. The main one is that it is difficult to determine whether you are sampling a temporal interval or a particular local environment. This is a distinction between reading the sequential layers of rocks as an evolutionary narrative of ancestors and descendants, or an aerial story where the sequential rock layers are preserving different biotas each with their own indigenous flora and fauna. In the 1981 movie *The Gods Must be Crazy*, an African Bushman finds a Coke bottle that fell out of an airplane. The Bushman has never seen anything like it and views the bottle as a magical object. The point is that if you think of material cultures as faunas and sample two environments, an "advanced" Western one, complete with airplanes and bottled soft drinks, and a traditional one of Bushmen who carry their beverages in emptied ostrich shells, you might come to the mistaken conclusion that because the Bushman's fauna was more primitive, it was therefore older, when, in fact, they exist contemporaneously.

Fossil faunas, for the same reason, are not that useful in determining absolute age, but they can be used to correlate relative age. If you find a certain group of fossils in one locality and then miles away find the same suite of species, it is likely that the fossils of the two localities are of the same age. Unfortunately, it is often difficult (especially in the case of rare animals like vertebrates) to find exactly the same suite of animals at different localities.

At a global level, this becomes even more problematic when one tries to make comparisons of faunas on different continents. Global comparisons are usually made on the basis of species that are similar, not the same. Even this approach, however, is rife with problems. Take Australia. Today (actually a few hundred years ago, before it became infested with feral cats, rabbits, and Europeans) the large animal fauna is predominately composed of marsupials like wombats, koalas, and bandicoots, which are considered primitive mammals. If everything on Earth today were exterminated and we only had the fossils, using the method of correlating similar organisms, a viable conclusion would be that the Australian fauna was more ancient than, for instance, the African fauna, because its members are more similar to extremely primitive animals that lived tens of million years ago. Consequently, correlations made on the basis of anything beyond the overlapping occurrence of the same exact species in two or more different localities are suspicious.

The rocks entombing the Jehol fossils are generally of two types: coarse-grained material that was deposited by small streams or rivers, and very finely grained ashy materials that were deposited at the bottom of ponds and lakes. Until very recently, nearly all of the fossils came from the fine-grained beds that preserved the spectacular feathered dinosaurs. These rocks are some times called paper shales, because each sedimentary lamination is as thin as a sheet of tissue paper. Such

The sediments of the paper shales are actually very fine laminations of rock, each indicating another depositional episode.

deposits simplify collecting, because the rocks break along the bedding plain. Many believe, although I don't, that fossils in a rock cause a zone of weakness that causes the slabs to split preferentially, just at the place where the fossils lie.

More recently, beautiful three-dimensional specimens of a variety of creatures have been discovered in the coarser fluvial sediments. These fluvial beds have even preserved animals that exhibit behavior. The best guess of how this happens is that the animal is buried under sediments so quickly that it is frozen in time. This is what I call the Pompeii model, and it can happen both with sand-mud flows, as has been documented in my Gobi Desert excavations where dinosaurs are recovered sitting atop their nests brooding their eggs, and in low-energy pyroclastic (volcanic ash) flows.

Mick Ellison's reconstruction of *Mei long* as a feathered animal in a sleeping or resting pose.

In the bitterly cold winter of 2004, the brilliant young paleontologist Xu Xing arrived in New York to start a two-year fellowship working in my research group at the American Museum. I have known Xu for years, all through his inexorable rise to his position as one of the most successful dinosaur hunters in the world and the greatest authority on the dinosaurs of Jehol. It took time for Xu to settle in, moving his family to an alien world. The first several weeks, he just tried to keep warm and to placate his wife. Xu had assured her that New York was warm, and heavy coats and down blankets were not needed. Not true this year, as the mercury didn't rise past freezing for what seemed like an eternity. After settling in, Xu showed me some new specimens of troodontid dinosaurs from a new locality near the village of Lujiatun. Xu Xing and I had described the first Jehol troodontid with Peter Makovicky, an ex-student of mine now at Chicago's Field Museum, in 2002. Troodontids are small coelurosaurian theropod dinosaurs that are very closely related to birds, and any new specimen is of interest. The game with these specimens is not, however, just the morphology of subtle bumps and measurements of bones. These guys had died suddenly and were preserved sleeping like the Roman victims of the eruption of Mount Vesuvius in Pompeii. Yes, sleeping. If you have ever seen birds asleep, you know they sit in a characteristic posture. Their wings lie on their sides and wrap around the body. The head is tucked in between the elbow and the torso. I couldn't believe that this specimen captured this position in death. This is just another piece of evidence showing how bird-like these dinosaurs are, not just in appearance, but in behavior as well. But how could these animals be preserved in such a pristine state, asleep?

Volcanic eruption usually connotes flaming rocks, oozing lava, and Krakatoa-like explosions. Yet other more silent killers also occur during volcanic events. One of the major killers in volcanic eruption leaves you not burned, but gasping for air. In 1986, a sulfurous gas cloud from a volcanic crater called Mount Chamberoen killed over a thousand people in the African country of Cameroon. Other deadly gasses are known to occur as parts of volcanic events worldwide. Because so many of these animals appear to be preserved sleeping or resting, gassing is a real, but hard to test, explanation.

Each rock unit that is thought to have been deposited continuously without significant gaps in sedimentation is called a formation. Formations can be composed of several types of rock. For instance the Yixian Formation has both fluvial rocks and paper shales. Several similar formations have produced the fossils that make up the

The lenticular brown layer is volcanic rock. Such volcanic rocks can be radiometrically dated and tell us that the underlying rocks are at least as old as the minimum date obtained in such an analysis.

Jehol Biota. The oldest of these is the Haifangou Formation, which is thought to be Middle Jurassic, or about 35 million years older than the rocks that produced the first feathered dinosaur *Sinosauropteryx*. Some think that this formation should not be considered a component of the Jehol due to its age and faunal differences.

Significant confusion exists as to what localities belong to what formations, how the formations relate to faunas, and how the different formations interrelate. This confusion is complicated by the extremely fractured and faulted landscape, a lack of good outcrops devoid of topsoil and plants so the underlying rocks can be directly seen, and more recent subterranean volcanism, which has cooked and altered many of the rocks.

Three formations have been classically recognized. Most of the great fossils have come from the Yixian Formation, which is further subdivided into a number of beds. It is a thick rock unit—up to 1.4 km thick—and probably represents about 10 million years worth of time. Above it is the younger Jiufotong Formation, which is also about a kilometer thick and represents about the same amount of time. Below the Yixian formation are several older poorly understood formations, like the Haifanguo Formation, which are much older, Middle Jurassic in age (about 160 million years old). These formations and their subdivisions are parts of a very complex three-dimensional puzzle. A puzzle which is missing many of its pieces, lost through erosion. What is apparent, however, is that the general environment and communities present in Northeastern China were very stable for a very long period. Yet understanding exactly how old the pieces of this puzzle are in absolute terms has been elusive.

Radiometric dating is the best way to determine exactly (within a calculated margin of error) how old a fossil is, or more precisely, the rock entombing a fossil. When the ash that fell to form Liaoning's Jehol deposits was ejected, it went from a high-pressure and temperature-liquid state to a solid one. When it was ejected from Earth, the molten liquid held a singular isotope of Potassium—Potassium 40. This was captured in the minerals that formed the ash cloud. Over time Potassium 40 decays into Argon—Argon 39. Because the rate of decomposition of isotopes is known (this is called its half-life), measuring the ratio of isotopes can determine precisely (within a small experimental error margin) how old the rock is in years.

I wish the radiometric dating of the Jehol deposits was this easy. The absolute dating of the Liaoning rocks is in fact a matter of contention, and different labs have produced different fossil ages. In some cases, dates as old as 147 million years have been reported, which would make the true birds from Liaoning the oldest ever unearthed, predating the current earliest bird *Archaeopteryx* (from the Late Jurassic deposits of Solnhofen, Germany) by several million years. On the other end of the spectrum, dates as young as 110 million years have been reported.

The basis for disagreement is both sociological and scientific, and it is often difficult to tease the two apart. Much is at stake here, in terms of nationalism, ego, and funding procurement. To have the oldest or first of something confers a certain amount of prestige. The scientific arguments about the age of the Liaoning deposits have several sources. There are problems inherent in radiometric dating itself. Different labs use different techniques, which sometimes makes it difficult to compare results. Problems also arise with false signals if the rock being tested has been reheated by more recent geological activity, which is the case with many of the Liaoning rocks, or if some of the Argon originally trapped in the rock was already Argon 39. To date, it has been possible to conduct radiometric testing on very few of the Jehol deposits. The majority of deposits have been dated using correlation methods. Some of these have even been tied to localities as far away as Western Europe.

Fossil plants, like this fern, are commonly found in the Liaoning deposits. Ferns were one of the major components of the under story vegetation in Jehol times.

In spite of all this, consensus is emerging concerning the age of the Liaoning faunas. Most of the evidence is now suggesting an Early Cretaceous age (roughly the 20 million-year period between 130 and 110 million years ago) for the Yixian and Jiufotang Formations—the uppermost and most fossiliferous part of the Jehol beds. These relatively recent dates conflict with the untrustworthy signals from correlation

The countryside near Sihetun is peaceful and green during the spring and early summer. The rest of the year it is a parched inferno or colorless frigid landscape.

because the closest relatives of many Jehol animals are found in deposits from much earlier time periods in other parts of the world. Noting these similarities, Zhexi Luo, a paleontologist at Pittsburgh's Carnegie Museum of Natural History, suggested in a 2002 *Nature* article that during the Early Cretaceous period this area of Asia acted as an isolated refugium, a sort of ancient Madagascar populated by primitive, relictual species. This hypothesis explains the presence of so many unique animals, called endemics, whose fossils are found here and nowhere else.

As is the case with Madagascar's lemurs and the finches of the Galapagos, with isolation comes specialization. While Northeast Asia was home in the Early Cretaceous to the last members of many groups of organisms, it appears to have been the spawning ground for the diversification of others. From the perspective of a biologist, faunas containing endemics and relicts are interesting for two primary reasons. One is that because endemics have evolved in isolation, they have existed in a sort of natural laboratory. Consequently, islands like Hawaii, the Galapagos, and the grossly understudied islands of Oceania are places where evolutionary biologists since the time of Charles Darwin have gone to carry out detailed studies of evolutionary patterns and processes. From my perspective, that of a systematist (one who is interested in deciphering the family tree of organisms), relictual species are crucial to a deeper understanding of life's family tree. Because they so often preserve extremely primitive characteristics (like Australia's egg-laying mammals, the echidna and the platypus), they are crucial elements to analysis of genealogical patterns and to understanding the sort of morphological transformations that living species have gone through in their long evolution.

Liaoning is not the most picturesque part of China. In fact, if it were not for the remarkable fossils of Jehol, I would never have traveled to the Northeast. But the endemic animals of Jehol have allowed Mick and me to sample the endemic culture of Liaoning. As they say: *"bai wen bu ru yi"* —seeing once is better than hearing a hundred times.

CHAPTER 3

Dragon Hunting

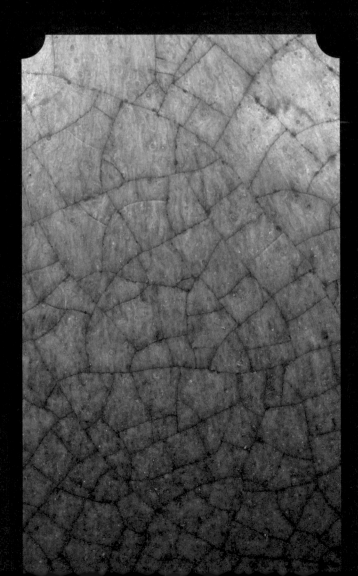

The Farmers Who Made Elvis

Unlike most fossils found today, only a tiny fraction of those excavated from the Liaoning deposits are collected by professional paleontologists. Instead they are "found" by local people. It is not as if the fossils are happened upon serendipitously in fields. Most are excavated in large quarry operations, both licit and clandestine.

Liaoning Province is poor. At the hub of China's military industrial engine, its cities are polluted, post-apocalyptic landscapes. The countryside is lush green and stifling hot for half of the year and brown, bleak, and cold during the other half. The farmers grow primarily sorghum, and the average annual income is less than $400. There are many issues here, and as China modernizes, the schism between rich and poor is even more apparent than it is in our own country.

One of the big problems for young men is the unavailability of wives. China's one-child-per-couple laws, combined with the Chinese cultural propensity for having sons especially in rural China (facilitated through all sorts of illegal practices like infanticide and sex-specific

A farmer tills his field on a steep hillside. The white areas above are sites of recent excavation for fossils.

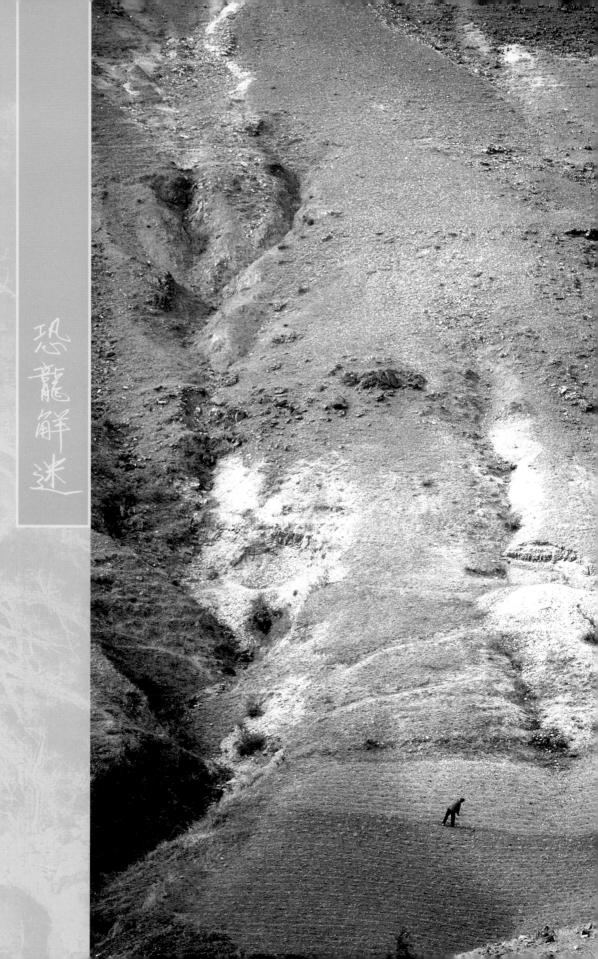

恐龍解迷

The main quarry at Sihetun. Here Mr. Li points to the layer which produced the first feathered dinosaur specimens. This entire excavation was made with the express purpose of fossil mining.

Li Dong Xue, the dautghter of Li Yinfang, with a handful of fossils from the main quarry near her home in Sihetun.

Gao Ke-Qin taking a break at the foot of the lion.

abortion) has resulted in a skewed sex ratio. Some estimates indicate that by 2020 there will be as many as 40 million more men of marrying age in China than women. Although better housing, televisions, and refrigerators are becoming more available in rural areas, even by Western standards they are not cheap, and the girls can afford to be choosy. A good fossil can bring $10,000 and even more, so finding one is like winning the lottery and getting the dream date at the same time—whether or not rules were broken in getting it. So a lot of rules designed to protect fossils are getting broken.

In the best-case scenario, a promising locality rich in well-preserved fossils is identified and subsequently excavated by local farmers working under the supervision of professional paleontologists. Unfortunately, only a few vertebrate specimens have been collected this way, by teams from China's Institute of Vertebrate Paleontology and Paleoanthropology, by field crews under the supervision of Gao Ke-Qin of Beijing University, and by Ji Qiang's guys.

In the worst-case scenario, local farmers untrained in professional excavation techniques surreptitiously remove fossils from quarries, risking damage to the specimens and providing little or no data as to provenance. Some of the quarries themselves are hidden, and there are even stories of tunnels with entrances hidden by shrubs that burrow into some of the richest layers. Because of the monetary value of the specimens, collectors are not always forthcoming with exact locality data. This has led to a lot of misinformation being passed around, as collectors try to do all they can to keep their excavations secret. Many of the collectors cannot resist preparing the fossils themselves, and they do a terrible job with substandard materials. Instead of removable adhesives, they use bondo and crazy glue. Their reasons for doing this are often just to see what they have, to determine how much a specimen is worth. In doing so, they usually decrease its scientific value.

Properly collected, many of the best specimens find their way into museums in China. When Li Yinfang found the *Sinosauropteryx* he packaged it as two separate deals. One slab went to Chen Pei Ji in Nanjing; the counterslab went to Ji Qiang at the National Geological Museum in Beijing. Li netted $750 for the slab that went to Ji Qiang's Geological Museum in 1996—a bargain basement price that would probably require two more zeroes today. The specimen in the collections of the National Geological Museum is the type specimen as it was described first by Ji Qiang in 1996. A complete study of the specimen requires a visit to both institutions. This is not an isolated practice.

In some cases, different scholars at different institutions have assigned two different names to the same species, based on different parts of the same specimen. This in fact happened when Gao Ke-Qin described a small long-necked aquatic reptile and named it *Hyphalosaurus lingyuanensis*, based on a single slab from the Yixian Formation. Unbeknownst to him, others were describing the counterslab, which they named *Sinohydrosaurus lingyuanensis*. Gao's name was published first, so it has priority.

The trade in illegal fossils remains lucrative and is thriving. Many important purloined specimens are smuggled out of China and sold openly on the international

Excavations made by local farmers in search of Jehol fossils. This is near the town of Chaoyang as it was in spring 2004.

market each year. Specimens have even been purchased by major international muse-ums, even though every single one was removed from China illegally.

Most legitimate museums have strict collection policies that forbid staff members from acquiring specimens that have been looted from their country of origin, even if their own national laws do not prohibit it. Yet several museums in Japan, the United States, and Europe have a more cavalier attitude. For example, Stefan Peters, former assistant director of the Senckenburg Museum in Frankfurt, was quoted in a 1998 *Science* article as saying, "It is better that museums acquire these specimens rather than some private collection." The Senckenburg Museum has acquired many important Chinese fossils for its collections and continues to do so. In the same *Science* news story, Japanese officials stated in defense of their purchase of Chinese fossils, "It is almost impossible for them to understand the details of

Chinese domestic law regarding fossils." This is patently false. Zhou Zhonghe stated, "It is more than clear that Chinese law forbids such exports of important vertebrate fossils."

In the United States, there has been some tightening up of the fossil trade. A specimen of *Confuciusornis* (a primitive avian), which was purchased by a museum trustee, was temporarily displayed at the New Mexico State Museum in Albuquerque. It was removed shortly after questions were raised about the legality of its export from China. There have also been domestic seizures of fossils, and they have been returned to China. Yet they were not seized because they were removed from China illegally, but rather because the import documents had been falsified, stating that the objects were building materials instead of fossil specimens.

The response of officials in the Chinese State Administration for Cultural Relics to the ongoing sanctioned importation of fossils by other countries that "this is robbery" is typical and to the point. Furthermore, a number of Chinese paleontologists have made a strong appeal that any institutions having such specimens should return them to China. A letter issued by the Chinese Society of Vertebrate Paleontology at the end of 2001 stated, "We hold the belief that a responsible museum should not buy and deposit any smuggled fossil, however justifiable it may seem" and "All conscientious scientists should not be involved in any trading or studying of the illegal specimens, however important they may be."

Although somewhat a gray area when it comes to legality, fossils are commonly sold throughout China.

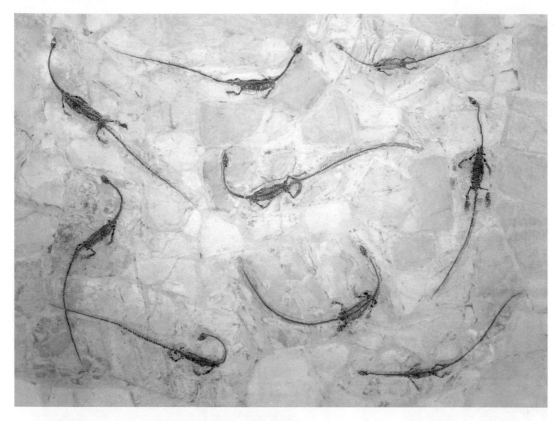

A nine dragon screen constructed by putting together 9 separate specimens of *Hyphalosaurus*. This one adorns the wall of a hotel dining room in Chaoyang.

It is a difficult situation. As a museum curator myself, I understand perfectly the impulse to acquire rare and important specimens. However, simply to disregard laws of another country because you either want a specimen or feel that you are saving it from something else is wrong. Lord Elgin hauled tons of priceless artifacts back to Victorian England from Greece with all the patronizing zeal of the age. We should know better. As long as Chinese paleontologists and officials deem that acquiring these specimens is illegal, it is our responsibility not to acquire them.

The brisk trade in spectacular and valuable fossils from Liaoning has also led to a cottage industry of fakery. This fakery is of three kinds: outright fakes, improvements, and chimeras. The outright fakes are pretty amusing. Here the aesthetic of the East parallels that of Tijuana. I have seen miniature *Tyrannosaurus rex*es, the London *Archaeopteryx*, a profile of Nefertiti, and even a likeness of Elvis, the King himself, complete with rhinestones embedded in the collar, skillfully carved into small curio pieces.

Improvements are another matter. Improvement has a long history in China. On a grand scale, the legendary limestone karsts on the Li River in Guangxi Zhuang were modified during Ming times, using gunpowder to fit the Imperial aesthetic. On a finer scale, the 18th century Qing emperor Qianlong had jadeite figures, originally fabricated more than 700 years earlier during the Tang dynasty, recarved to fit the fashion of the day. So it is unsurprising that some enterprising fossil dealers cannot resist making specimens just a little bit better by carving new details or painting on feathers. A decorative example is the wall of the main restaurant in the Central Hotel in Chaoyang, where nine different *Hyphalosaurus* specimens have been annealed together to form a screen. Traditionally, such screens were meant to ward off evil spirits. The Chinese regard nine as the largest numeral (with the same pronunciation as "forever") and the dragon as an auspicious beast.

Far more dangerous are the chimeras. Some are comical, like the snake formed by the vertebral column of several fossil fishes attached end to end, or the parrot dinosaur (*Psittacosaurus*), sporting an artfully attached coat of hair made with fine fish bones. The most infamous chimera was *Archaeoraptor*.

This bizarre creation graced the pages of *National Geographic*'s November 1999 issue. Looking like a hung-over rooster with teeth and a tail, *Archaeoraptor* was proclaimed by Christopher Sloan, one of the magazine's senior assistant editors and the author of the article, to be the "true missing link in the complex chain that connects dinosaurs to birds." This was the middle bit in a very unfortunate chapter in paleontology.

In 1999, a specimen was smuggled out of China and offered for sale at the Tucson Gem and Mineral Show, the world's pre-eminent fossil sales event. There, it was purchased for the Dinosaur Museum, a private institution in Blanding, Utah, by M. Dale Slade, the museum's president, founder, and patron. He paid $80,000 for the smuggled specimen. The Dinosaur Museum is the brainchild of Steven and Sylvia Czerkas, who are prominent dinosaur artists with no formal training in vertebrate paleontology.

To formally name a dinosaur, or any other species, a description must be prepared and published. For important and high-profile specimens like this one, such descriptions often appear in top-ranked international scientific journals. The two most prestigious are *Science* and *Nature*. The Czerkases assembled a scientific team composed of Philip Currie, Tim Rowe (the acknowledged world expert on CAT scanning of fossils), and Xu Xing. A plan was laid whereby the specimen would be laundered back to a Chinese institution sometime after a paper describing *Archaeoraptor* was published. In this way, the Czerkas et al. team would avoid criticism that they worked on a dirty illegally imported fossil. During the spring of 1999, a paper was prepared for *Nature* on the specimen. Because the suite of preserved characters on the *Archaeoraptor* specimen (a very birdlike front part of the body and a very primitive typical theropod-looking back part) would make this an unbelievable discovery, *National Geographic* committed itself to publishing a popular nonscientific article on the specimen that was scheduled to appear coincident with a paper in the primary scientific literature.

Unfortunately for the National Geographic Society, the project was shrouded in secrecy. Doubly unfortunate for Czerkas et al. (or maybe fortunately, for some of the authors involved), the paper was rejected by *Nature*. Czerkas than turned it around to *Science*, where it was again rejected. Peer review, and pre-peer review, had rejected the paper's conclusions and evidence, and it never appeared in a scientific journal. That left *National Geographic* as the journal of record, because it had developed its secret project betting that *Science* or *Nature* would accept the paper. When they didn't, *National Geographic* did not have enough lag time to pull the article. Its editors announced the "great discovery" with much hoopla and media attention at a national press conference.

Many in the scientific community, including some of the authors of the rejected paper, were already suspicious. Tim Rowe had noticed while conducting a three-dimensional tomography study that the contact between the back end of the skeleton and the rest of the body was problematic. Even in pictures that I saw of the specimen, there were issues. The two hind feet were part and counterpart mirror images of the same foot, slapped together on opposite sides of the animal. *National Geographic* started to take a lot of abuse, and Bill Allen, the senior editor at *Geographic*, was incensed. Things really got worse when in December Xu Xing notified *National Geographic* in an e-mail that the counterslab to the tail was found in a private collection connected to a very different animal, the small dromaeosaur *Microraptor*. In January 2000, *National Geographic* admitted that the specimen of *Archaeoraptor* was a chimera. *Archaeoraptor* was an elaborate fake. Instead of a missing link, it was simply pieces of at least two different fossil animals cobbled together to deceive people and drive up the price. The august National Geographic Society is rarely so embarrassed, and the embarrassment extended to many others involved. Philip Currie summed it up when he referred to the whole episode as "one of those great sinking feelings one hopes never to have in one's career."

The infamous *Archaeoraptor* chimera.

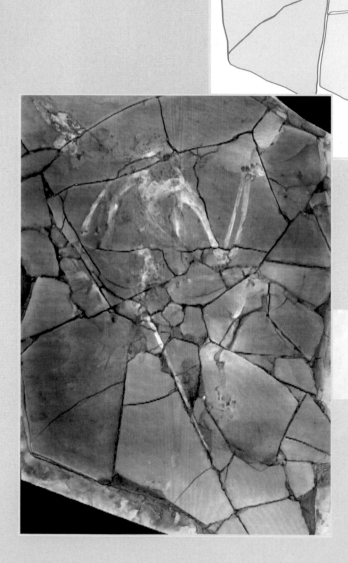

Definitive evidence that "*Archaeoraptor*" is a fake. On the left is a CT scan and on the right a diagram of the specimen. The individual colors correspond to different specimens from which the composite was constructed.

Further analysis by Tim Rowe, using CAT scans, conclusively revealed that the specimen had actually been assembled with 88 pieces from several different fossils. He published the result in a *Nature* paper in 2001, with a detailed color diagram of how the specimen was crafted. The main parts are the tail of the *Microraptor* and the body and skull of a previously unknown avian, which was named *Yanornis* in a 2002 *Nature* paper titled "*Archaeoraptor's* better half."

National Geographic asked me to join an internal investigatory panel of experts to report on the *Archaeoraptor* fiasco. The meeting was held at the Smithsonian, and because the National Museum of Natural History has a strict collections policy (similar to my own museum), the Czerkases were required to relinquish title of the specimen to the Chinese government before it could be brought into the museum to be examined by our panel. Storrs Olson, Smithsonian bird curator, made sure that a dirty specimen was not allowed on his premises.

Our inspection of the specimen quickly confirmed what we already suspected. Lewis Simons, who won a 1986 Pulitzer prize for international reporting, sat in on our panel meeting. He concluded in his article, published in the October 2002 *National Geographic*, that the entire affair was "a tale of misguided secrecy and misplaced confidence, of rampant egos clashing, self-aggrandizement, wishful thinking, naïve assumptions, human error, stubbornness, manipulation, backbiting, lying, corruption, and, most of all, abysmal communication."

Unfortunately, the *Archaeoraptor* affair has been jumped on by creationists to cast doubt on the authenticity of all the Liaoning fossils, the dinosaurian origin of birds, and even the tenets of evolution. Charles Colson, of Watergate fame, reporting for the Cornerstone Church Online, asserted, "What was supposed to be startling news has turned out to be yet one more example of the scientific community peddling fraud as scientific fact."

While the reaction of creationists might have been expected, some scientists fomented surprisingly similar rhetoric. In an attempt to discredit all the feathered dinosaurs from Liaoning because they conflict with his preconceived theories, University of North Carolina ornithologist Alan Feduccia went so far as to claim in a February 2003 *Discover* piece that "*Archaeoraptor* is just the tip of the iceberg. There are scores of fake fossils out there, and they have cast a dark shadow over the whole field." The common thread connecting creationists and some ornithologists is that they ignore the fact that the process works. No peer-reviewed scientific journal accepted the claims for *Archaeoraptor*, and it was scientists themselves who exposed the fraud.

A fossil store in the village of Lingyuan. Such large dealers have a sophisticated network of contacts in the countryside and some operate on both sides of the law.

Changes are happening in Liaoning that will, it is hoped, suppress the illegal export and smuggling of fossils. Anti-corruption efforts are aimed at dismantling the complex web of officials who turn a blind eye to these illicit activities. Some individuals have even been arrested. In 2003, two men were sentenced to 10 years in prison and fined $24,000 for trafficking in fossils from Liaoning to Korea. Recognizing that the famous fossils may be a revenue source, new small museums have sprung up in Dalian, Beipiao, and other cities. Some of them are sending people abroad to study international curatorial standards and are beginning to apply them locally. Often these institutions employ local farmers to collect specimens. Legislation at both the provincial and national levels has resulted in the creation of the Liaoning Protection Bureau and made it slightly easier and more legitimate for farmers to come forward with their specimens and be lawfully reimbursed for their efforts.

But problems still exist. One of the government's main attempts to quell the tide is to sanction a few official fossil shops where specimens can be evaluated for scientific importance. Yet, other fossil shops operate quite openly. For instance in the dusty wild East town of Lingyuan, Mick and I visited a fossil shop in the spring of 2004. There, selling quite openly, were some of the finest Liaoning specimens that we had ever seen. The proprietor of the shop explained that he had some better ones in a "safe place," presumably because they were too good, or maybe just as an insurance policy.

The summer of 2004 saw some bizarre happenings, where, as reported in *Nature*, in an effort to protect important fossil sites from looters, the Liaoning Bureau of Land Resources blew up a series of important fossil sites. Furthermore, several farmers who were collecting from the sites were detained and threatened with arrest. The local people and some Chinese scientists saw this as an attempt at extortion, where the fossil dealers and their middlemen were upset with the prices they were asking. Unfortunately, a particularly valuable specimen appears to have been crushed by being pulverized with a tractor. The local farmers, according to the *Nature* article, would rather destroy it than turn it over to local officials they believe to be corrupt.

But more and more is being done to help preserve these valuable treasures. In China, there are real efforts, both at national and local levels, to restrict the illegal trade, but still compensate farmers equitably when an important specimen is found. International societies (such as the Society of Vertebrate Paleontology and the Society of Avian Paleontology and Evolution) have developed ethics statements that criticize the smuggling and those who facilitate it. While it's not a perfect world, and it is still pretty crazy out there, it seems that a modicum of law is coming to the old East.

CHAPTER 4

古老的庭院，熟悉的虫子

Ancient Gardens, Familiar Bugs

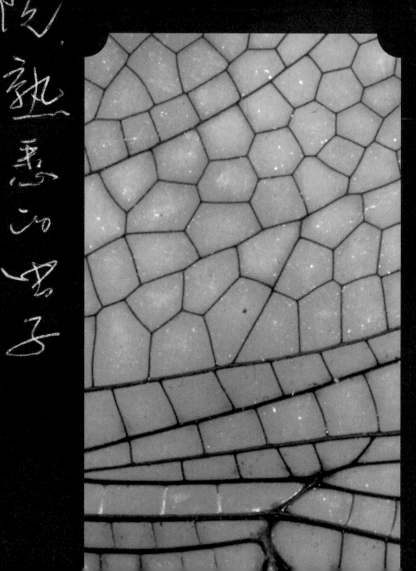

And an Advanced Course in Southern Chinese Food

There aren't just dragon hunters in Liaoning. Plants, insects, arachnids, mollusks, crustaceans, and the fossil remains of a host of other organisms, even worms, have been found in the Liaoning deposits. They make up the vast majority of fossils discovered there. Some, like the showy flower-like inflorescences of cycads, look just like modern flowers, but they're not. The flora is extensive, and conifers, ferns, and horsetails vegetated the shorelines of the lakes and rivers of the region during the Early Cretaceous. We are not merely getting an intimate look at the feathered dinosaurs, but the habitat, the ecosystem, the world that once was theirs. The rich detail of the landscape that is emerging is unprecedented in paleontology. So much diversity is captured in the fossil record of Liaoning that we can reconstruct the Jehol habitat with greater efficacy than any other dinosaur habitat in the world. The world that is created during this exercise is one that contains a mix of the familiar, the recognizable unusual-looking, and the just plain weird.

The extinct *Ephemeropsis* is probably the most common fossil in the Yixian Formation. Sometimes thousands are preserved on single slabs. The specimen shown here is an aquatic larva of an adult winged form.

An orb weaving spider. The variations in tone on the legs and body correspond to patterning in life.

Moving from land to water, the thickets, leaf litter and littoral habitats were home to lots of bugs, that is to say, all terrestrial arthropods (Arthropoda is the group or phylum that includes all insects, crabs, shrimp, spiders, and the like). The fossils found at Liaoning include scores of species, most of which are only beginning to be studied, consequently, they are, as yet, unnamed. Many are familiar, including ants, dragonflies, cicadas, cockroaches, spiders, and beetles. Indeed, arthropod fossils are the most common animals found in the fine-grained thinly laminated paper shales. Many scientists feel that the major diversification of bugs into the general types of insects that we see today occurred about the same time as the first dinosaurs and mammals appeared, about 235 million years ago. So by Liaoning time we would expect to find a nearly modern fauna, and we do find many modern types of insects.

There are, however, still some differences between the Liaoning insect fauna and the insect fauna of today. The ways the mix of fossil arthropods differs are that some modern things are missing, and some Liaoning animals are today extinct. Many modern sorts of bugs that fly, crawl, and crunch beneath your feet are not found as

Although these look a lot like flowers, these inflorescences actually belong to a cycad.

fossils in the Liaoning deposits. Insects like butterflies, ants, and honeybees either hadn't evolved yet, or were very rare and not diverse. But fossils of more primitive members of the groups to which they belong (Lepidoptera: moths and butterflies, Hymenoptera: stinging insects like bees, wasps, and ants) are known from the Jehol Biota. As the diversification of the major insect groups is often tied to the origin of flowering plants, it is apparent that in the Liaoning fauna we are seeing the beginning of the diversification of the modern insect fauna.

Organisms that live in aquatic environments have a high probability of being fossilized in the paper shales, as these fine-grained sediments were laid down on the bottoms of lakes. Therefore, a plentitude of aquatic fauna and flora is preserved. Animals associated with water include a rich array of familiar invertebrates—like those that would occupy any mid-latitude pond, trout stream, or salt marsh today. Dragonflies skimmed the surface, dodging diaphanous booby traps set by orb-weaving spiders; back-swimming beetles and predaceous mayfly larvae patrolled the mid-water; and the lakebeds were home to crayfish, snails, and clams. Jehol was even home to the Yao Ming of water striders, as some specimens measure nearly 15 cm across.

Crayfish are farmed today throughout China. They are not uniformly popular, but in Shanghai, Nanjing, and Hangzhou, crayfish restaurants abound. A red crayfish icon decal is applied to the establishments' windows. These aren't native crayfish though; the farmed ones originate in North America. Yet fossil crayfish are numerous in the Jehol sediments. The fossil crayfish belong to a still-living group that is distributed nearly worldwide. The living examples are remnants of an ancient group little changed for hundreds of millions of years. If the crayfish family tree that includes these animals is superimposed on a map of the planet that is about 275 million years old, their distribution (at least up until the advent of recent global homogenization due to commercial farming) can easily be explained by the dynamic nature of continents. Since the 1960s, it has been known that the continental land masses (or continental crust) essentially float like pond scum on heavier, denser rocks. These denser rocks are called oceanic

This spectacular wing looks like a butterfly, but it is not. It is the wing of a scorpionfly. We have no evidence that butterflies had evolved at the time of the Jehol deposits. Many of the insects of the Jehol have wings that when found by themselves resemble butterflies to the non-specialist.

crust, and they make up the ocean basins on the planet's surface today. With large-scale movements of the material underlying the crust (the mantle), the continental bodies move around, collide, form composites, and divide. This is a slow process, but it can be measured in real time. Accurate laser satellite sensors have determined that the Atlantic Ocean is widening, pushing Europe and North America apart by about 2.5 centimeters a year. Remarkably, much of Sumatra shifted nearly 100 meters in the catastrophic 2004 Boxing Day earthquake.

Geologists have provided a detailed understanding of continental rearrangement. In broad strokes, an ancient continent called Pangaea broke up about 235 million years ago (at the same time the earliest dinosaurs are found) into a northern continent called Laurasia and a southern one called Gondwana. This was followed by the separation of Laurasia into North America and Eurasia. Gondwana fragmented, and India rafted across the proto-Indian Ocean to finally collide and anneal with Eurasia, a collision that produced the Himalayas. Finally, in Cretaceous times, about 80 million years ago, Australia and Antarctica separated from Afro-South America and concomitantly, Afro-South America and Australo-Antarctica fragmented, forming the continental masses of the contemporary world.

Water striders were very common in Jehol times. This one is normal, meaning the same size as those alive today. However, some giants, up to 14 centimeters, have been found.

Like crayfish, the distributions of many other animals and plants reflect ancient topological arrangements of continents and oceans. Even before the advent of plate tectonics (the body of theory developed in the 1960s that explains drifting continents), distribution of organisms was used as evidence for continental connection.

Ostracods are strange arthropods. They are like little shrimp, composed mainly of large heads, with their entire bodies covered by a pair of calcareous shells. They have small appendages, are filter feeders, and can thrive almost anywhere, from the vast open ocean to the muddy bottom of a freshwater lake. Some can even live in soggy moss near the water's edge. They are very diverse, with more than 10,000 living species, and relatively unchanged in the more than 500 million years since their first appearance in the fossil record. The fossilized shells, or valves, are really common in some Liaoning sediments and look like thumbprints on many of the slabs.

Because they are so common as fossils, they have been used as tools to correlate among rock units. Some of the early attempts at this suggested that some of the Jehol rock units that contain avians were

The Jehol crayfish are part of a group with a worldwide distribution. Here this specimen nicely shows what is meant by slab and counterslab.

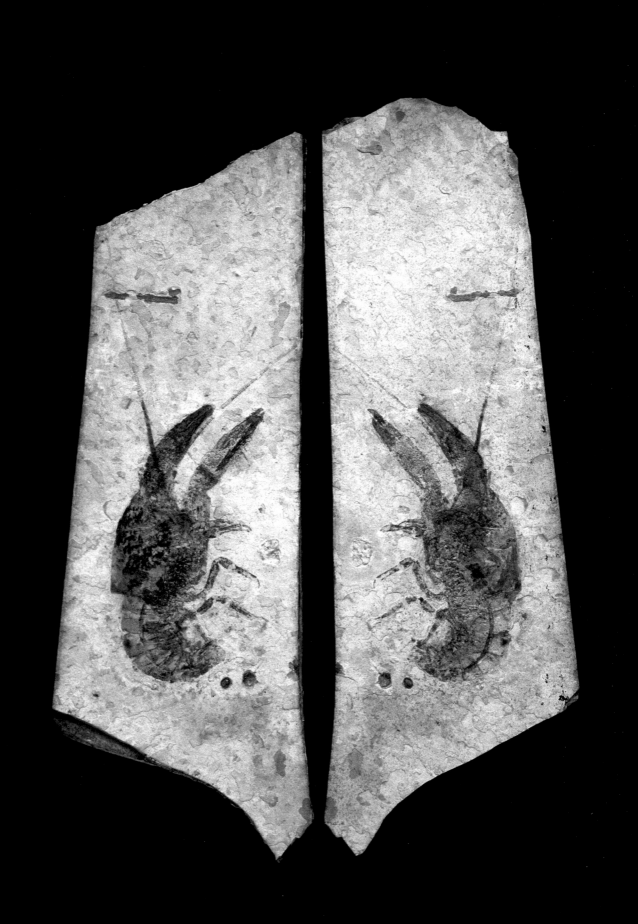

older than or as old as the rocks of Solnhofen in Bavaria, which have produced *Archaeopteryx*. Under scrutiny, however, the ostracods just all look too much alike for them to be of any use for the fine temporal subdivisions that are needed. Looking similar are conchostracans, the fossils studied by Chen Pei Ji. The large live ones around today are tasty menu items when cooked in a broth of soy sauce and rice wine, a conclusion reached only after exhaustive study.

If there is a place in China that begins to resemble how parts of Liaoning would have looked 120 million years ago, it is the highlands of Indochina, the province of Yunnan, near the border with Vietnam, Laos, and Burma. Nowhere else is the environmental and ethnic heterogeneity of China more apparent. Northern China looks and feels gray. Northerners eat steamed bread, noodles, and salty boiled peanuts. The people are wonderful, but I do not know anyone who would characterize them (like New Englanders) as particularly warm, outgoing, or fun-loving.

In the South, in the area between Kunming (Yunnan's capital city), in the foothills of the Himalayas, adjacent to the Burmese border, the thick but semi-arid rhododendron forest is rich in wildlife. Given the locals' omnivorous proclivities (meaning eating everything and anything), such abundance might not be expected. It was in southern China where SARS is thought to have jumped into humans when locals ate infected Palm Civet meat. Here are brightly colored tropical birds, elephants that roam the forests, and a variety of ethnic minorities, each with its own indigenous language, dress, and cuisine. Watercourse valleys dissect the undulating topography of low hills, alternately terraced and forested. There are a few large lakes, the shores of which teem with aquaculture operations producing fish and shrimp.

In Yunnan Province alone, there are more than 20 distinct ethnic and cultural groups, including Tibetans, Yao, Naxi, the Muslim Hui, and Indochinese hill tribes like the Wa, Bai, Miao, and Dai. Each group wears distinctive, brightly colored, and heavily embroidered clothes, a disappearing but still active tradition. Many of the groups are dispersed, so cultural incongruities abound. When you're hungry and traveling in Yunnan, you never know what to expect.

We were on a trip to Lufeng to a site where many dinosaur fossils older than the Liaoning specimens have been found. It was here that indigenous Chinese dinosaur paleontology really started, with Yang Zhungjian's discovery of *Lufengosaurus* in 1938. Several regional museums in this area house important fossils, crucial to our understanding of global dinosaurs. Because we had some free time in Beijing, we

The fossiliferous beds of southern China exposed above the city of Lufeng in Yunnan. It was in these rocks that the first dinosaur specimens from China were found.

arranged a visit to Southern China to see some of these important specimens. On the drive southward from Kunming, Mick and I stopped at a place by the roadside that served food. It was a diner.

Our colleague and traveling companion, Guan Jian, told us we were in a Naxi village. The Naxi are descended from Tibetan nomads with a unique pictograph-based written language. Naxi society is matrilineal: Women own the property, control the money, and pretty much run the show. In some areas the old tradition of "walking marriage," in which men live in their parents' home and only visit their wives' houses, is still practiced. Relationships are informal, children live with and are supported by their mothers, and there is no recognized paternity.

The room in the diner was dark and the seats low. We were hungry and road-weary. Jian's Beijing *hua*, the Mandarin dialect spoken in and around China's capital, is not well understood by many people in the south, such as the ones working in this eatery. After a lot of difficulty, pointing, and laughing, we ordered the food. We slunk onto the stools in the zone for cool beverages. Suddenly, a hand appeared across Mick's face, wrenching his head to the side. Noting our debilitated state, or at

A vendor peddling animal parts used in traditional medicine on the streets of Lufeng.

Jian's urging, the proprietress had commanded the waitresses to give us "chair massages," a combination of *qi gong* (a kind of Chinese massage), shiatsu, chiropractics, and wrestling. For the next 30 minutes, as we slurped sweet rice wine from rough sections of green bamboo whacked with a machete into cups, two Naxi girls alternated between pressing their bodies against our backs and cracking every bone in our necks, shoulders, arms, and hands.

A table full of brightly colored food with a variety of textures and shapes was placed before us: fried beetle larvae mixed with chilies; giant water bugs; black, red, and green rice steamed in whole bamboo sections; and the *pièce de résistance*—whole grilled Chinese bamboo rat. We didn't know what would be produced next. Perhaps, like the gingko, other thought-to-be-extinct creatures (maybe a feathered tyrannosaur) would be offered up as the next course. We were lucky; it was time for the floor show.

Stuffed with things we had never eaten before, we loaded into the car and began the last leg of our journey to Lufeng. Riding through the lush countryside, content and nearly comatose, I found it easy to imagine what Liaoning might have looked like during the Early Cretaceous. It was a world rich in color and teeming with life, the lakes and rivers filled with myriad fishes and aquatic reptiles, the sky populated by birds, dragonflies, and pterosaurs, and the land the habitat of dinosaurs and mammals.

The fossil plants we have found enable us to reconstruct the forests of the feathered dinosaurs with a degree of certainty. What is apparent is

Brightly colored embroidery against a dark field is characteristic attire of many of the Yunnan minorities.

A tasty plate of fried insects. Lunch in Yunnan.

that modern sorts of flowering plants were minor players in the Liaoning biota. Relatives of modern types of nonflowering plants made up most of the bushland. But it is still tricky drawing a complete picture of the plant-life—unlike animals, plants make lousy fossils. Plants today are primarily angiosperms, and their very hard wood makes them great candidates for fossilization. This hard wood is thought to have originated as a response to aridity sometime during their history, probably long after the Liaoning animals had vanished. The somewhat softer and pulpier trees of the Liaoning forest are very poorly preserved, usually only as a carbonized film in the paper shales.

Rarely do we find an entire plant as a fossil. Nevertheless, plant parts like leaves, cones, seeds, and small stems are often abundant. It is very hard to put these back together into a single plant. For instance, think about how a pine tree, with the same size and shape of cones and needles, may be a dwarf naturally occurring bonsai variety with a gnarled and twisting trunk low to the ground, or a hundred-foot monster straight enough for a ship mast or telephone pole.

The fossil plants found in the Jehol sediments are primarily ones with small leaves or needles. Broad-leafed plants (like many tropical plants) lose water quickly through their leaves; consequently, small leaves and needles are good indicators of fairly arid conditions. The dominant types of trees in the forest were conifers, similar to today's pines and firs. Although related to modern conifers, these trees looked a little different. From the fossils that we have, we know they were not gargantuan; instead, they had trunks with a maximum of about 45 cm. Living conifers usually have their needles arranged in groupings of two or three. The Liaoning fossils show a more crowded arrangement, with as many as six to nine needles emanating from a single bud. Other conifers are similar to broad-leaf conifers; rare today, they are present in Japan and Australia.

The remaining big trees were gingkos. Gingkos are gymnosperms, and the modern trees are the last remnants of an extensive radiation of their kind that began over 150 million years ago. Recent Western studies indicate that gingko is a potent source of anti-oxidants, and

A seed-bearing cone. Such three dimensional plant fossils are rarely found, as are reproductive structures like this one attached to the rest of the plant.

Conifers were the dominant trees in Jehol times. They differ from today's in that there are more needles converging on a single place than we see on trees in forests now.

some research, although controversial, has suggested that gingko improves brain function and memory. It has even been suggested as a remedy for early onset Alzheimer's disease. Apparently at the end of the last ice age, gingko habitat was pitifully tiny, so tiny that there are no natural populations. But for nearly 4,000 years, gingko has been used medicinally in China for everything from increasing stamina to treating impotence, and gingko trees were cultivated in Buddhist monasteries in the Southeast. About 1,200 years ago, they were taken to Japan, where they were discovered by the German botanist Engelbert Kaempfer. Like the coelacanth, gingkos were known as fossils before they were found alive.

The fossil gingkos of Jehol were almost identical to today's gingkos, except that the characteristic broad fan-shaped leaves were dissected, giving the leaf a much smaller surface area, again a sign of aridity. Gingkos are bisexual, and city dwellers know them for the vomit-like smell of the fruits which encase an edible seed. Gingkos line the sidewalk of the 77th Street entrance to the American Museum of Natural History. Each winter, I see pedestrians checking the soles of their feet to try to account for the smell (77th Street, after all, is a main artery to the dog runs of Central Park).

Because angiosperms were still rare, the understory vegetation was primarily ferns, superficially similar to today's bracken ferns and cycads. But these were not the barrel-shaped thorny cycads of Southern California yards and golf courses. Rather, they were delicate segmented plants, kind of like poinsettias, with soft stem segments and whorls of leaves occurring at intervals. The needles, leaves, and stems of all of the Liaoning trees, bushes, and shrubs were, if they are like their living descendants, notoriously hard to digest. This would make for a lot of leaf litter on the floor of the forest, as well as associated mosses, horsetails, and other small plants.

Although Liaoning has produced one of the greatest diversities of fossil organisms of any site in the world, tantalizing evidence suggests that much more remains to be found. This is no more apparent than in the case of the co-evolution of plants and insects. In December 1997, Sun Ge, a paleontologist at the Nanjing Institute now at Jilin University in Changchun, received a newspaper-wrapped package of fossils. Such packages, usually from Liaoning farmers hoping to make a sale, often appear on the desks of Chinese paleontologists. These, however, had come from one of Sun Ge's colleagues. Recently returned from Beipiao, he thought that the specimens were ferns, common fossils in the Jehol deposits. Sun Ge placed the specimens in a drawer, only to pull them out for study a few days later. Under a microscope, Sun Ge saw something that made his day. Most of the fossils from the Jehol are gymnosperms like those just described. These have small leaves, and this specimen looked pretty typical. But under the microscope, Sun Ge saw something different. The leaves were actually small seeds arranged helically. He quickly realized they were the remains of an angiosperm, a flowering plant, one of the oldest and most primitive ever discovered. He named it *Archaeofructus*. Since then, many more specimens and species of *Archaeofructus* have appeared. Characteristics of the flowers, which are small and unisexual, and the strongly dissected leaves indicated that it was an aquatic plant. Other early flowering plants have appeared as well, including a terrestrial one, *Sinocarpus*, that is apparently much more advanced than *Archaeofructus*. Other tantalizing fossils may be angiosperms, but because of the problems with plant fossils described earlier (we just find pieces, not many entire plants) we can't tell yet if they are really the precursors to today's flowering plants.

Fossil and living gingko leaves. The only difference is that the fossils are deeply dissected, probably an indication of a more arid habitat. In the fall gingkos turn brilliant yellow and red, so it is no doubt that the Jehol forest was colorful.

The segmented and branched stalk of what some consider an early terrestrial angiosperm.

Small ferns are common as fossils. Today ferns hold only slight nutritional value, so they were probably not a big food source for herbivores like *Psittacosaurus*.

The presence of *Archaeofructus* and *Sinocarpus* would seem to indicate that we are sampling the dawn of the evolution of flowering plants. These angiosperms did not have spectacular, eye-popping flowers and would hardly elicit a second look today. Showy flowers act as attractors for insect pollinators, and the lack of them in *Archaeofructus* (as well as other aspects of its anatomy) suggest that this plant was aquatic, pollinated by the action of water.

Other Cretaceous fossils from Liaoning, however, suggest that flowering plants were not restricted to an aquatic environment and were far more diverse than the current fossil record chronicles. Throughout their history, plants and insects have had a mingled destiny. In today's world, insects are pests, pollinators, parasites, guardians (as in the case of aphid-eating beetle larvae), and mimics of plants. Often, these interactions are species-specific and so intertwined that the existence of both plant and insect depends on the presence of the other.

Some orchids require certain insects to pollinate them. This is accomplished by specialized mouth parts of the insect exactly fitting into the nectar-producing structures of the plants. In the process of feeding, pollinating insects transfer pollen from flower to flower. Without the specialized mouthparts, bug feeding would be as frustrating as trying to plug an American

appliance into a Chinese electrical out-
let. The mechanism explaining these
intricate similarities is called co-evolution, a process where two (or occasionally more) species change in concert
through time.

Co-evolutionary arrangements may
be part of the dynamic that has fueled
the tremendous diversification of
insects and flowering plants. There is
some evidence in the fossil record to
support this, because the tremendous
diversification of pollinating insects
occurs at the same time as the diversification of flowering plants.

Brachycercan flies, like this one that was described by Ren
Dong, are important indicators there were terrestrial
angiosperms these flies pollinated that have remained thus
far undiscovered as fossils.

Knowledge of specific relationships between contemporary plants and insects
allows us to make predictions about their early diversification, based on a few sparse
fossils. A particularly telling example of this is the case of brachycercan flies, two
species of which have been found at Liaoning. This group of flies still has living rep-
resentatives, including flower, bee, horse, and robber flies. Modern brachycercan flies
are characterized by long tubular proboscises, mouth parts modified for sucking nectar out of flowers. Like other insects that subsist on nectar, these flies are active pollinators and very specific as to what kinds of flowers they feed on. Although until
now, no terrestrial flowers, aside from perhaps *Sinocarpus*, have been found in the
Jehol fossil deposits, the fact that the fossil brachycercan flies have the same nectar-
feeding structures as their modern descendants strongly indicates that there were
several kinds of terrestrial flowers in Northeastern China during the Early
Cretaceous. If and when such fossils are unearthed at Liaoning, they will be some of
the earliest true flowers ever discovered.

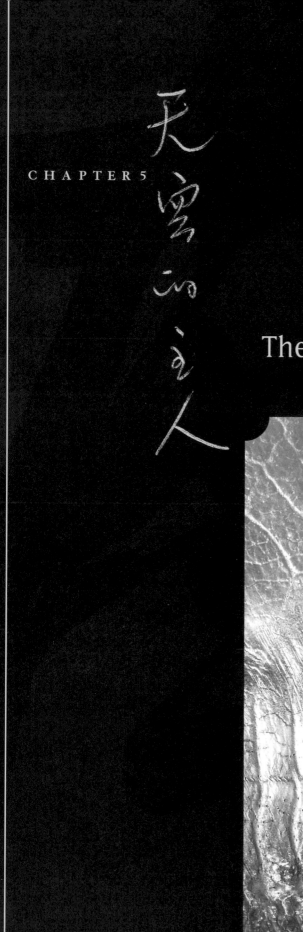

CHAPTER 5

The Masters of the Sky

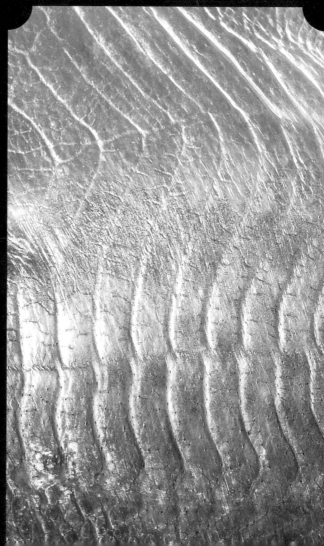

And the Creatures That Crawled, Swam, and Were Eaten Below

One early morning, Mick and I walked through the market on the way to see Ji Qiang—it was Day One of another trip—we were in typical first-day form. Ji had e-mailed me that he had a new pterosaur, and we were eager to get to his office. We detoured on the way; a little walk might make us feel better.

Frantic bicycle trucks and loud motorcycles loaded with produce blasted through crowds of shoppers. Because household refrigeration is relatively new and locals like their food extremely fresh, most Chinese shop every day. The market's frantic pace has a rhythm of its own, to which shoppers move seamlessly, crowding around the freshest or cheapest produce. Or so you would think. Baskets of live fish explode from too many fish in too weak a basket. Dying fish take their last flop on the filthy floor, spewing fish slime on everyone within a 10-foot radius. Chattering rubber-booted fishmongers pursue the slippery fugitives, even in the heat of battle never dropping a single ash from the cigarettes dangling from their mouths. This is, of course, the funniest thing the crowd has ever witnessed.

Markets in China serve up almost anything one
would desire, dream, or imagine.

Asian markets are among my favorite places. The people, the colors, and the smells—especially the smells—overpower and seduce. The smells aren't bad, but compared with the antiseptic markets at home, even the blind can identify what is for sale. And everything is for sale here. It is sensory overload. If you are what you eat, that explains the rich and varied culture of Asia and the Chinese.

The market is a PETA nightmare. The living, the dying, and the long dead pack tanks. Animals are caged in baskets; they are tethered, bailed, or hung from meat hooks. Goats, sheep, pigs, cows, dogs, and even cats—this is all food. Food is hauled from the oceans and rivers or trapped or shot in the forests. Even with my expensive education in biology, I continue to find critters I can't identify. Yet there is order here. Generally, things that live in water—fish, frogs, turtles, and snakes—are all in one place. Snakes are noteworthy because they inevitably escape. I remember one hilarious occasion when Dong Zhiming, the legendary IVPP dinosaur hunter, took me to a snake restaurant in northeast Beijing. This was the early 1990s, when there were far fewer foreigners around. The restaurant was filled with flimsy aquariums holding every sort of imaginable serpent. Some looked like visions of hell, with snakes so thick you could not tell where one started and another began. Suddenly, we heard shattering glass, and the entire contents of one aquarium created a waterfall of serpents onto the floor. The patrons went crazy, some standing on their chairs, others laughing as the waiters and waitresses attempted to capture the slithering dinner items. I'm also wary of snakes, having had one too many cobras waved in my face during late-night drinking sessions. The amount of snake bile-infused baijiu Mick has consumed may be the source of his legendary virility. Or maybe not, but it may help answer the question of why there are a billion Chinese.

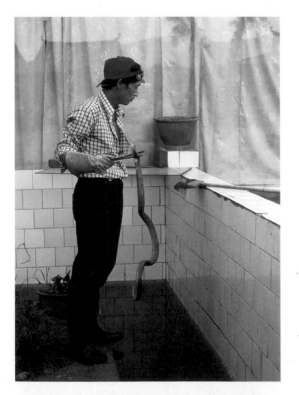

A waiter picks a cobra in preparation for snake baijiu followed by a snake meal.

Things that fly—birds and the occasional bat—are kept in another section. Mammals, caged or flayed, are down another aisle.

Imagine a visit to an antediluvian market, where the animals that were tethered, put in aquariums, or garroted were those of Liaoning 130 million years ago. How different would it be?

During the Early Cretaceous, the dominant land animals, at least in size and ubiquity, were dinosaurs, not mammals, as they are today. Granted, while a *Velociraptor* tethered to a meat hook would draw attention even in a Chinese market, the rest of the foodstuff would look just about the same.

Any great Chinese meal includes fish. Fish are the most

The short necked champsosaur *Monjurosuchus* preserved on a slab with fish remains. Perhaps this represents a mass death assemblage caused by a volcanic event.

common backboned fossil from Liaoning, and they were the first vertebrate fossils from these beds to be described. Most of these fish are *Lycoptera*, and sometimes thousands are preserved in a single layer, indicating some sort of mass mortality. *Lycoptera* is a small herring-like fish that is common in fossil deposits throughout the world. Before Liaoning's Jehol deposits were radiometrically dated, the presence of *Lycoptera* was used as an important piece of correlative evidence for determining the age of these formations. Simply put, the beds were considered to be Early Cretaceous because *Lycoptera* were also found in Early Cretaceous beds in other parts of the world.

Lycoptera are far from the most spectacular fish found in these sediments. Not only are there more fish fossils, there are more kinds of fish than any other vertebrate in these deposits. Even though I am totally uninterested in fish paleontology

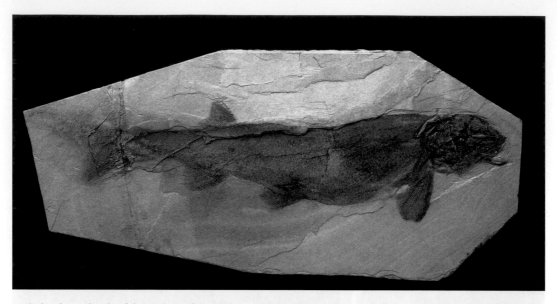

Both archaic and modern fishes are known from the Liaoning sediments. This specimen is a relative of the bowfin, which is still alive today. Scientifically important, such specimens are also aesthetically beautiful.

(too many bones, and the animals are too much like their modern counterparts), I like fossils. So when looking through various collections of Liaoning fossils, it is easy for me to get stuck admiring the fish. There are so many amazing specimens that an entire volume could be devoted just to them.

Many of the fish are the sorts of primitive examples one might expect. There are large sturgeons and paddlefish. These fish have cartilaginous skeletons, very small scales and upturned tails—all remnants of a fish diversity that was present before the diversification of more modern types of bony fishes. Relatives of these fish, although nearing extinction because they are a tasty food item, are still to be found in the Yangtze and Yellow Rivers. There are also large armored fish, which are all extinct today, but they are distantly related to the modern bowfin.

Generic fish fossils are flat monochromatic carcasses. Those from Liaoning aren't. Often having preserved remnants of color and pattern, they look like ghosts, in the sense that the bones are visible, but so are stripes, spots, and outlines of bodies. In some examples, even the shading of the eyes is apparent. These fossils look like someone started with a fish skeleton and painted stripes and spots, using the stone as a canvas.

The presence of pattern and color on many of the Liaoning fossils is one of their most spectacular attributes. A detailed understanding of how this is preserved is still to come; however, a little is known. Usually the patterns and colors that we observe on the specimens are not the actual colors of the animal in life. What we are seeing is a film laid down by bacterial decomposition of different pigments in the scales, skin, feathers, or internal organs of the animal. So far, most of these decomposition products are a uniform dark gray color. Yet we have every reason to believe that the Jehol world was not black and white, but just as colorful as today's.

Other kinds of small cold-blooded vertebrates inhabited Liaoning's rivers and ponds and their shorelines during the Cretaceous. While most would not look appetizing to most Americans, Chinese culture has a rich tradition of eating reptiles and amphibians. Consequently, all sorts of snakes, frogs, salamanders, and turtles are found in markets. As reported in the *South China Morning Post*, one enterprising restaurateur tried to drum up business during the 2003 SARS panic by serving an endangered species of salamander (under special license from the government). According to the chef, "it is especially good for women, because it will give you a clear complexion."

Fossil salamanders from this part of China span a significant amount of time. The earliest ones are specimens from Late Jurassic rocks that underlie the rocks that preserve the Jehol Biota. They are several million years older and are the earliest true salamander fossils to be discovered. Fossil salamanders are also found in the Yixian Formation, one of the rock units that preserve the Jehol Biota. Like today's salamanders, all of these early ones went through a process of metamorphosis from fish-like juveniles to salamander-like adults. All stages of their life cycle are preserved so finely that sometimes a salamander's large fragile gills lie as dark films next to its body.

Discovering primitive members of modern groups is one of paleontology's greatest quests. Salamanders are exciting landmarks in this quest, because they have an interesting distribution, obscure origins, and a fascinating developmental cycle that has made them model organisms for modern biologists. Frogs, the snake-like caecilians, and salamanders constitute the living diversity. The picture is murky when it comes to the origins of living amphibians. All these animals have degenerate, highly specialized skeletons, making their bones difficult to compare with those of other groups of animals, living or dead. Over the past five years, thousands of fossil salamander skeletons have been collected by teams led by Gao Ke-Qin. These are some of the oldest salamander fossils ever discovered, and they have been used as evidence that the origin and early evolution of salamanders occurred in proto-Asia.

Because these animals are so primitive, only one can be assigned to a still-living group of salamanders, called the Cryptobranchidae. Cryptobranchids are one of the

Part formality, part entertainment, and always overload for eyes and taste buds, banquet meals are an important part of China's cultural fabric.

two most primitive living salamanders, and these fossils are the most primitive members of this group ever discovered. Cryptobranchids today include the largest of all salamanders, the Japanese giant salamander, which can grow to nearly two meters long. They have big heads and large eye sockets, which led to the misidentification of a European fossil cryptobranchid in 1725 as a human killed in Noah's flood.

Because the skeletons of these fossil salamanders are less degenerate than those of their modern descendants, they can be more easily compared with even more ancient and primitive amphibian groups, some of which are also preserved in larval stages. We have just begun to study the paleontology of Northern China, and the salamander studies reflect this. There is much here for Gao to study, with big implications for refining our ideas about the origins and evolutionary geography of modern amphibians. Because salamanders are such an ancient group, their current distributions may be highly influenced by the rearrangement of continents through plate tectonics. A better understanding of salamander interrelationships may provide us with additional data concerning the timing and pattern of the continental movements described earlier.

Unlike salamanders, fossil frogs (the most diverse group of terrestrial tetrapods aside from birds today) and lizards are not at all common. Perhaps because lizards don't spend much time in the water, they have less chance of being preserved. One of the better lizard fossils was found in the thoracic cavity (presumably the last meal) of the first specimen of the fuzzy dinosaur *Sinosauropteryx*. The other lizard specimens are small, nondescript animals known from poor fossils, only two of which, *Dalinghosaurus* and *Yabeinosaurus*, have been named. While the relationships of *Dalinghosaurus* await further study, it is evident that *Yabeinosaurus* is a relative of advanced modern lizards like geckos and skinks.

Today, frogs are plentiful and easily farmed, and they taste good. Called country chicken, frogs of all sorts are found as food items throughout China. My favorite way to eat them is broiled on a stick in the night market. They can be raised in small pools and when found in natural settings they are often abundant. That should make them prime candidates for fossilization, but very few fossil frogs have been found. Not even tadpoles are common. Furthermore, like the lizards, the frog fossils that are found are not the sort of archaic forms we might expect or wish for, to inform us about the origins and relationships of modern groups.

We know that frogs are very old, because frog-like animals that are more closely related to frogs than anything else have been found as fossils in Madagascar in

A fossil of the enigmatic frog *Mesophryne* from the Yixian Formation south of Beipiao.

250 million-year-old rocks and in slightly younger rocks in Poland. The Jehol frogs are limited to just a few species. The best preserved of these, *Callobatrachus*, is a medium-sized frog, about 15 cm long. But there are other forms as well, like the enigmatic form *Mesophryne*, a frog that is so unusual that it can not be assigned to any known group of frogs. *Callobatrachus, however,* is related to a group of frogs called discoglossids. Discoglossid frogs are present in Asia, Europe, and North Africa. Before the discovery of *Callobatrachus*, no Asian fossil discoglossid had been found. Hence, the historical distribution of discoglossids in Asia can now be traced back at least to the Early Cretaceous, about 130 million years ago.

Curiously, no crocodiles have been found, even though crocodiles are ubiquitous in similar-aged deposits worldwide. On the small end of the crocodile scale, their ecological equivalents may have been the short-necked champsosaurs. Champsosaurs are a superficially crocodile-like group of reptiles, whose exact relationship to other

Hyphalosaurus is a common fossil represented by thousands of specimens. It was a weird looking animal with big eyes on a small head with a ridiculously long neck and tail.

reptile groups is not well understood. Also enigmatic is why they are the only major group of reptiles to have survived the terminal Cretaceous extinction event (about 65 million years ago) and then become extinct during the Cenozoic Era (the so-called Age of Mammals).

Champsosaurs were the first group of tetrapods described from the Liaoning beds. These species (of which the first specimens to be discovered were lost during World War II) were described by Japanese paleontologists during the Imperial Japanese occupation of Liaoning. Most Jehol champsosaurs are small, about a foot or so long, and Gao Ke-Qin, Ji Qiang, and I have worked on several of the new specimens. Some of these, like *Monjurosuchus*, are so well-preserved that they show the animal had webbed feet and keeled scales. Large champsosaurs are less common, though many undescribed specimens exist in Chinese collections.

My favorite champsosaur is *Hyphalosaurus*, because its fossils are aesthetically beautiful and huge in number, which will allow for important research down the road. This extinct reptile must have lived in pigeon-like abundance, because so many specimens have been unearthed. Often, many are found on the same slab. Like so many other Jehol cases, this is a sign of a catastrophe. There are so many specimens that *Hyphalosaurus* is the most common non-fish fossil souvenir on the Chinese curio market. Each comes neatly framed on blue or red velvet, usually with a small plaque incorrectly identifying it as "Swimming Dinosaur" or sometimes "Swuming Dinosaur" or other permutations of the same letters. They can be seen everywhere tourists shop in China, on eBay, and even in Chinatown curio shops.

Unlike most champsosaurs, *Hyphalosaurus* is not crocodile-like. Instead, it has a short body, a long tail, and a preposterously long neck, ending in a tiny head with huge eyes. It looks like one of those friendly monsters Dr. Seuss might have created.

Many readers of popular science over the past couple of decades are familiar with ideas about how changes in developmental patterns may hold clues to the way organisms evolve. The origin of novel characteristics upon which natural selection can operate has always been a key question for evolutionary biologists. There have always been issues about how one feature transforms into another, such as how a pentadactyl (or more probably a limb with more than five digits) transformed during evolution from a fin. The developmental process provides us not only with the end morphologies (things that early on look like fins, and, in the case of larval salamanders, are fins, that modify into fingered limbs).

In development we can see not just the end products but transformation among these endpoints. What then if we could reshuffle the timing, onset, and developmental rate of structures in the growing organism? This would provide organisms with a huge pool of variation on which natural selection could later sort. This is an old idea, originally proposed by such 19th century naturalists as Karl von Baer, Louis Agassiz, and Ernst Haeckel and recently promoted by the late Stephen Jay Gould of Harvard University and by Keith Stewart Thomson, director of the Oxford University Museum of Natural History and my Ph.D. adviser. Yet to understand rearrangements, we need to first document changes in developing animals. Some of these changes are simply differential growth of various components of the creature. Documentation of this is called allometry.

To do an allometric study, an extensive growth series is required. While this can be observed in living animals, such a series is usually not present in the fossil record. *Hyphalosaurus* is an exception. Fossils coming out of Liaoning seem to reveal a complete sequence of *Hyphalosaurus* development from juvenile to adult. If we take even a cursory look at this series, it is obvious that something is going on with the neck. In *Hyphalosaurus* babies, the neck is about the same length as the body; in more mature animals, it is longer; and in adults it is preposterously long. Why? Basically, the neck extension during growth occurs at a faster rate than the growth of the trunk. Just like puppies that grow into their oversized feet, baby *Hyphalosaurus* necks grew faster than their bodies.

Understanding allometric relationships between parts of organisms, if combined with a family tree, can lead to an understanding of how these relationships

A tiny juvenile *Hyphalosaurus*. These fossils are among the most common tetrapods in the Jehol fauna.

have changed over the course of evolution. Changes in allometry usually have to do with timing, that is, the onset of development of a particular body part, or the rate of developmental changes in relation to some other part of the body. This is called heterochrony.

A good example of heterochrony is in salamanders, where some kinds of salamanders never appear to completely metamorphose into adults. Instead, they reproduce while still looking like larvae, in the sense that they have gills and are completely aquatic. This is because maturation of their sex organs occurs before maturation of the body into an adult terrestrial form with no gills. Before the family tree of salamanders was known, these forms were thought to be the most primitive groups of salamanders.

Now we know, however, that the salamanders that reproduce while looking like larvae are actually descended from forms with a fully terrestrial habitus, and through evolution, the timing of development of the sex organs has just been shifted. A carefully documented allometric study of the skeletons of all champsosaurs, especially in relation to a family tree, may provide important clues to the generation of novelty in this group and may explain the process by which long-necked and short-necked forms appeared.

Turtles have a special place in Chinese culture. They are relished as food and are used as a yin tonic in Chinese medicine. Unfortunately, their long lives, and the ease with which they can be caught, have made the future of many species incompatible with a burgeoning and upwardly mobile Asian population. In every market where we travel are boxes, aquariums, and pens full of turtles, some of them extremely rare species. It is estimated that several million turtles a year cross China's porous southern borders, destined for the markets and restaurants of Guangzhou, Beijing, and other cities. Much of the living turtle diversity of Asia will disappear before it has a chance to be adequately studied.

In spite of the carnage, in my own mind turtles are tied to their central place in one of the favorite Chinese insults. "*Wan ba dan*" (turtle egg) is a very bad thing to call someone. On several occasions Mick and I have spent hours inquiring why calling someone a turtle egg is such a rank insult. We are no nearer the truth because, after gallons of Tsing Tao, our brothers inevitably begin contradicting and insulting each other. Usually it starts with Mick asking why *wan ba dan* is bad. Bao will answer, "because the turtle lay egg underground and ...," only to be interrupted by Makai, who tells us, "Bao is a very stupid man. He knows nothing and is full of shit." Bao then counters with, "Makai is a %$#@!," etc., and the conversation degenerates into a battle of

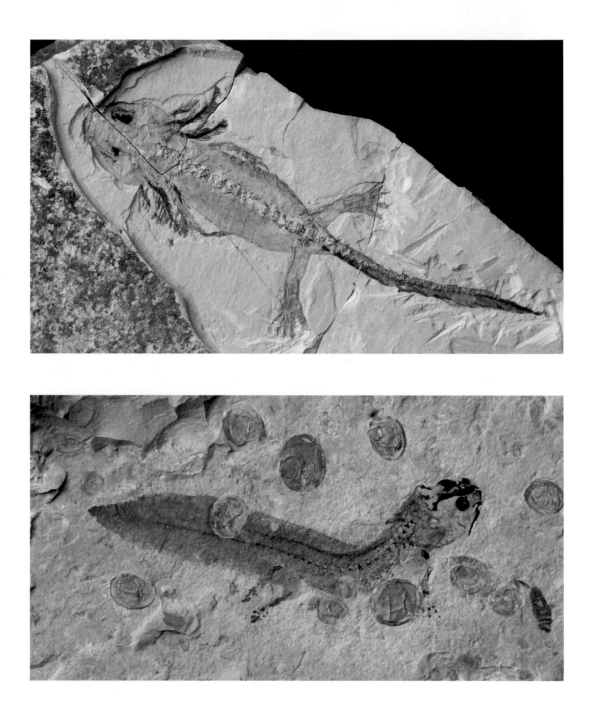

Salamander fossils are found throughout the Jehol and in earlier rocks regionally. Many have been found—everything from finned juveniles to adults. These fossils provide us with important clues about salamander origins and relationships.

A man who enjoys life—the connoisseur and overall renaissance guy Makai.

boorish shouting. As far as we have been able to divine, the insult is rooted in chelonian copulation habits. Because of the architecture of the turtle body, the female turtle cannot see who is screwing her. Consequently, the turtle egg does not know who its father is, and it's a colorful term for bastard, we think.

Fossil turtles are commonly found in Chinese markets. We have even seen them for sale in apothecaries and in the sort of shops that sell crystals and other *feng shui* items. Little is known of the early history of turtles. Unlike most modern animals that are significantly different from their extinct ancestors, turtles are always recognizable as turtles. There is no *Archaeopteryx* or Lucy of turtles. When they first appear in the fossil record (about 230 million years ago, long before Jehol times), they are almost modern in form. The only difference is that they could not pull their heads back into their shells. Perhaps the females of these primitive forms could see who was fathering their eggs.

Fossil turtles that are found in the Liaoning deposits look no different from the ones swimming in restaurant tanks. There are two kinds: soft-shell turtles, which are related to extant soft-shell turtles, and nanhsiungchelyids. The nanhsiungchelyids have lumpy shells, huge noses (the largest in all turtledom), long tails, and a very hard-to-pronounce name coined after the locality where these animals were first found. They resemble modern-day snapping turtles but are unrelated. Occasionally, especially in the lower Liaoning beds, mass burials of turtles are unearthed. They represent small ponds that dried up, condensing all of their inhabitants, first in a shallow puddle, then in a mud hole that completely evaporated, dehydrating and killing all that could not walk away. While this sounds rare, such accumulations are often encountered in the fossil record and can even be observed happening today in places like the Florida Everglades and the plains of East Africa.

The only noisy nonhumans in a Chinese market are birds, and there are lots of them. There are quails, many varieties of chickens (in Indochina you occasionally even see fighting chickens, locally called boxing roosters), pigeons and doves, and ducks—lots of ducks. As with most animals in China, every part of the duck is con-

ducks—lots of ducks. As with most animals in China, every part of the duck is consumed, and parts that are discarded elsewhere, such as the tongue and the webbed feet, have been elevated to delicacies.

The birds of Liaoning are fascinating, but the masters of the Jehol skies were a diversity of pterosaurs. And that is why, that early morning, we had set out to find Ji again, now ensconced in his new office at the Chinese Academy of Geological Sciences. Pterosaur fossils have been known for more than 200 years, especially from the classic Solnhoffen deposits of Southern Germany and from the south coast of England. Some of the first bird fossils collected were misidentified as those of pterosaurs, simply because they had long gracile bones that formed fragile wings. Pterosaurs, however, were not closely related to birds, nor were they flying dinosaurs. They were flying reptiles whose evolutionary relationships to other reptiles unique to the Mesozoic, such as dinosaurs, plesiosaurs, and ichthyosaurs, are still being sorted out, but they are usually considered close dinosaur relatives.

Just like birds and bats, pterosaurs evolved flight independently. Each solved the problem of wing structure differently. The wings of birds are composed largely of the humerus and the radius and ulna (the upper arm and forearm). Interestingly, though bats are related more distantly to pterosaurs than pterosaurs are to birds, their wing structure is superficially more similar. In bats, the wing is composed of a relatively short arm and an enormous hand, whose fingers form the support of the wing. Pterosaurs had a wing that was supported along most of its leading edge by a single, very elongated fourth finger that supported a skin membrane.

Some European and Brazilian fossil beds have produced pterosaur specimens with preserved impressions of the wing membrane that suggest it was naked and leathery. Accumulating evidence, however, suggests something different from a simple leathery flap. First, a poorly known pterosaur from Central Asia seems to be covered, at least in part, by hair-like structures. Because these specimens are not well-preserved, the function and distribution of the structures on the body have not been accurately determined. Still, there was enough evidence to suggest that pterosaurs were not flying lizards. Second, a closer examination of some well-preserved pterosaurs from a number of locations reveals that their wings were reinforced by internal structures called actinofibrils. Actinofibrils were oriented so that when the wing was outstretched, it opened like a Chinese fan. In flight, these structures stiffened the wing, operating like the battens on the sail of a boat, and were important in organizing the wing during folding.

Today's turtles are little changed from those that lived in the Early Cretaceous, over 125 million years ago. Somehow an adult *Hyphalosaurus* and this baby turtle met their fates together.

Some pterosaurs were very large. Here are the teeth and jaws of a specimen with a wingspan over 2 meters.

Pterosaurs are often found in the Jehol sediments, but, due to their extremely fragile skeletons, are not often well-preserved. Nevertheless, many kinds have been described. While most are fairly small pigeon-sized forms (such as *Dendrorhynchoides*), some, such as *Sinopterus*, have large albatross-sized wingspans, ridiculously small bodies, and grotesquely flat heads with airfoils on them. Some of these specimens preserve enough soft tissue to conclusively show that pterosaurs had fuzz. Until recently, none of them, however, unlike the case of *Sordes pilosa*, an amazing pterosaur from Tajikistan, is so well-preserved that it shows how the integumentary (e.g. fuzzy) covering was dispersed across a pterosaur's body. Was it present on the wings, just the body, or both? Did these animals have fuzzy heads and legs?

Ji Qiang stood in the morning light, smiling. He never understood why it was so hard to find him. It was April 2001, the beginning of spring in Beijing. Ji's new operation was very impressive, compared with his old digs at the National Geological Museum. The only weird thing, perhaps auspicious, was that the sidewalk in front of the building was covered with bird shit under a couple of trees. Ji explained that this was the site of an old temple, and the birds still returned there to roost. Inside, instead of cold basements and cavernous gray rooms with an unstable electrical supply, his new offices and labs were bright and neatly appointed with new desks and chairs, as well as the latest in computer equipment. As usual, we exchanged pleasantries, laughed a lot, Mick and Ji smoked, and I drank tea. After the usual banter, Ji grabbed a small, green satin box from the table adjacent to his desk. Quickly he opened it and handed it over, only to reveal the most remarkable pterosaur specimen I had ever seen.

It was small, about the size of a swallow. Classified within the genus *Dendrorynchoides*, it is a member of a really strange group of anuragnathid pterosaurs, small forms with broad, almost frog-like heads and sparse, peg-like teeth. The animal had sharp claws, as the sheaths that cover the bony underclaw were preserved. The sheaths doubled the apparent size of the claw and ended as thin points. All of this would make this specimen a great fossil, perhaps the best-preserved anuragnathid ever unearthed. Yet what astonished me was the body covering. This animal was a flying fuzz ball. "That is way cool," Mick announced.

Close inspection revealed several different kinds of soft tissue. The wing membrane extended from wrist and finger to ankle. Membrane was preserved between the hind limbs. This structure (called an uropatagium) is also found in most bats, which use it to stabilize flight and catch insects. And both the wing and the uropatagium were reinforced by radiating, subparallel rows of actinofibrils. All of this indicated that the animal was incapable of bipedal locomotion, a style of walking that had been previously suggested for pterosaurs. But what I saw first was the "fur."

I had to be certain that the hair-like structures all over its body were not fake, and that they were not intermixed remains of plants or some other animal. The possibility that they were actinofibrils or some other internal structure also needed to be entertained. It did not take long to be convinced that what I was seeing was real: The entire pterosaur was covered with these structures, and they were external. The length of the fiber varies, depending on where it is on the body, and each fiber thins from its base to its tip.

Over the next couple of days, we were able to establish that the "hairs" were distributed on both the pterosaur's body and wings, with particularly bushy clusters on

This pterosaur specimen, about the size of a swallow, is notable for the preservation of actinofibrils which worked like battens in the wing and for its fluffy body covering. This animal's entire body was covered with filaments. While not from the Jehol *sensu stricto*, it was found in older Jehol-like beds which spill over into Inner Mongolia Autonomous region. Related pterosaurs are found in Liaoning.

This fluffy pterosaur had widely spaced tiny teeth. There are indications of a fibrous integumentary covering around its mouth similar to some modern birds.

its neck and tail. Small fibers also lined its mouth and were similar to the bristles around the mouths of mammals or the beaks of flycatchers.

What was the purpose of pterosaur fuzz? It is hard to answer this with any sort of certainty, because the work is just beginning. It is clear that the primary use for integumentary fibers that cover the entire body in animals that are alive today is insulation. Such insulation is a requirement for warm-blooded animals, which generate metabolic heat. Usually these are highly active animals like birds and mammals. It is obvious that the hair on bats serves primarily as insulation, but this is also the true for the feathers of birds. Feathers on flightless birds have no aerodynamic function, yet they have been retained to maintain metabolic body heat.

We do not have a thermometer that can measure a pterosaur's temperature, but the presence of a body covering that may have functioned as an insulator is highly suggestive. An even more perplexing question is how these structures relate to feathers. Could this mean that a fluffy body covering is primitive for both pterosaurs and dinosaurs? One day at lunch, Ji summed it up: "Maybe we're all wrong; maybe all the Jehol dinosaurs were fuzzy."

Popular opinion has it that we live in the age of mammals. When most people think of animals, they think of elephants and dogs. When people, especially

Americans, think of food, they think of steaks and hamburgers. The Chinese opinion is that pork is the king of meats. Today, in the age of mammals, mammal meat is at the top of the food chain. A few years ago, wild game restaurants were the rage at home in New York. In an attempt to cater to exotic tastes, menus were expanded to include venison, wild boar, and the occasional beefalo, bison, and antelope. This is tame, compared with what goes on in Asia. Many of the animals we have talked about so far can be found in contemporary Chinese markets. Mammals are also there. Everything from common supermarket stuff like pigs and cows to Bronx Zoo material like pangolins and ferret badgers, awaits shoppers in the market—on the hoof or sliced to your satisfaction.

Few people realize that mammals are as ancient as dinosaurs. Both first appeared in the Triassic about 230 million years ago. For the first 70 million years of their history, however, our mammal ancestors did not diversify much. They were almost all mouse-sized seed and bug eaters. Many think they were nocturnal. Various ideas, notably the stifling competition from dinosaurs, have been proposed to explain this, but there is little in the way of real evidence, except for the fact that the evolution of our kind didn't really get started until the nonavian dinosaurs disappeared.

The mammals of Liaoning were small forms, which many would argue are not really mammals at all, because they are only distantly related to modern mammal groups. Over half a dozen different kinds are known, and most of them were unspecialized forms with teeth suited for a diet of insects and other small animals. Fossil mammals are among the rarest in the Liaoning quarries, although a few spectacular three-dimensional specimens of near-mammal relatives have recently been unearthed in fluvial rock units near the bottom of the formations that produced the Jehol Biota.

Unlike the younger rocks, which were laid down by volcanic ash at the bottom of the ponds, these rocks were deposited by the action of running water. Earlier, we talked about how the velocity of water influences the quality of fossils that are deposited. Still water usually preserves better fossils. But there is a flip side. Stream and river-laid rocks (called fluvial deposits), do not often produce specimens that preserve soft tissue and only rarely preserve small articulated skeletons. The Jehol specimens are, therefore, of two types: the slightly younger, well-preserved but squashed flat specimens from the fine-grained paper shales, and the slightly older, but usually larger and incomplete, specimens from the fluvial beds.

The larger mammals from fluvial beds are called triconodonts and are very primitive. Most of the animals that lived during the Mesozoic were small shrew-like crea-

Meng Jin, a colleague at the American Museum, holding the skull of *Repenomamus*.

tures, a few a little bigger—about the size of an opossum. When my colleague Meng Jin here at the American Museum told me that he had just brought back a big fossil mammal from China, I didn't think anything of it. I was expecting something maybe cat-sized. When I dropped by his office, I saw a skull lying on his table in protective padding. I thought it was something from the Cenozoic, the era after dinosaur dominance, as its diagnostic bits were hidden by packing. Then he showed it to me. It was huge, almost 25 centimeters long. Called *Repenomamus,* this animal was as large as a very big badger.

Initial preparation on one specimen showed a small skeleton in its abdominal cavity. Was this evidence that these were live-bearing animals? Unfortunately not, because the skeletons turned out to be juvenile psittacosaurs, common elements of the Jehol biota, instead of a triconodont baby. *Repenomamus* was probably right up there with the protofeathered tyrannosaur *Dilong* as the *Psittacosaurus'* worst nightmare.

By Liaoning standards, most mammal specimens from the paper shales are not that impressive. They look like diminutive flat rats: generic rodents trampled paper thin, preserved as thin smudges of guts and bones surrounded by a spectral aura of hair covalently bonded to a subway platform floor.

One of the Liaoning fossil "mammals" preserves impressions of hair. It is not surprising that these sorts of primitive mammals or near-mammal relatives were hairy.

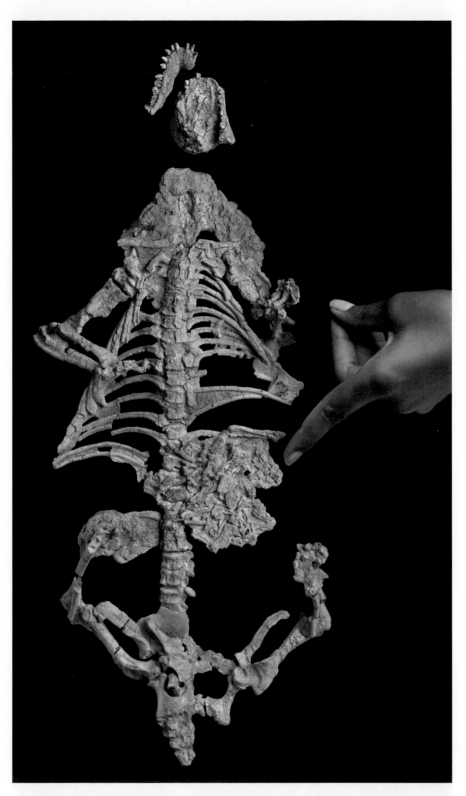

A skeleton of *Repenomamus* with its last meal of baby dinosaurs in its stomach.

The Jehol mammal *Maotherium*. This mouse sized animal preserves hair and resembles a flat rat on a New York street.

On the one hand, it underscores just how far back in vertebrate history such structures as hair, feathers, and other integumentary coverings like those of pterosaurs are present, and on the other, just how modern in appearance Jehol animals were.

Another primitive mammal, *Eomaia*, further highlights the modern aspect of Liaoning

The ubiquitous *Lycoptera*. These small fishes were some of the first fossils to be studied from Liaoning and are sometimes preserved hundreds to a slab.

animals. Living mammals are divided into three groups: mammals that lay eggs like platypuses; marsupials that carry their undeveloped young in a pouch like kangaroos and opossums; and placentals. Placentals are mammals that carry their young internally, attached to their mothers via a placenta from which the young feed until they are more or less fully developed. We and most other living mammals are placentals. *Eomaia* is a placental mammal, denoting that this group diverged very early, a view supported by some analyses of molecular DNA sequences. While this shows that mammals of essentially modern guise inhabited the Jehol landscape, it begs the question why it took 50 million years, until after the demise of traditional dinosaurs, for them to begin to diversify into the forms and types we see today.

How well-stocked is our antediluvian market? I venture to suggest that if we could blast back millions of years ago and efficiently harvest the Jehol ecosystem and cage and package it, there would be plenty to eat. The turtles, fish, the occasional frog, and rat-like mammals, or their modern equivalents, have been wolfed down by Mick and me at virtually every opportunity in the course of our travels. The furry pterosaurs would look like weird Stella Lunas, and the champsosaurs like, well, something you have never seen, but there's always something surprising at the market.

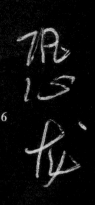

CHAPTER 6

Kong Long

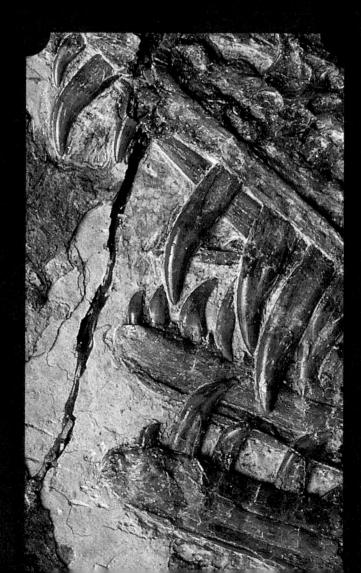

Chinese Medicine, Pajama Fashion, and Big Bones

In Imperial times the emperor and his court escaped the Forbidden City in the summer by retreating to the Garden of Clear Ripples, 20 kilometers from the city center. Built during the reign of Qianlong on the site of an ancient palace, construction of the Summer Palace, as it has come to be known, represented an expensive and impressive technological achievement. Large lakes were excavated and canals dug, the fill being used for false hills. In 1860, during the Second Opium War, Qianlong's Summer Palace was destroyed by Anglo-French armies in retribution for the government's closure of the opium trade.

It was rebuilt on the same spot by the Empress Dowager Cixi 30 years later, using funds earmarked for the reconstruction of the Chinese Navy, only to be severely damaged and looted again by the Eight Power Allied forces during the Boxer Rebellion of 1900. This was brutal business, as can be evinced by Kaiser Wilhelm II's order to troops that they "make the name German remembered in China for a thousand years so that no Chinaman will ever again dare to even squint at a German." The site

Chinese apothecary stores are sources of everything animal vegetable and mineral. Remedies mixed up by the pharmacists represent an ancient tradition based on special properties of natural materials.

Traditional medicine old and new.

Digging for dragon bones has a long history, and these miners provided some of the first fossils to end up in western collections.

suffered further depredation when many of its greatest treasures were stolen by Kuo-Min Tang soldiers and hauled off to Taiwan, where many were sold into private European and American collections. Today, after extensive remodeling and restoration, the site is the largest Imperial garden still in existence and draws millions of visitors a year.

Mick was among those tourists on one hot summer Beijing day, where the air was so thick it made a steam room seem arid. Nearly drowning in his own sweat, he bought a bottle of water from one of the touts, ripped off the cap, and downed most of it with a few gulps. Things went bad for the next nine days.

Traditional Chinese apothecaries are magical places. To Westerners, they resemble 19th century natural history museums, or a scene from a Harry Potter film. Lining the walls are cabinets with what seems like thousands of drawers, each labeled with a character. Dried seahorses and pipefish in jars line the shelves; sacks of desiccated snakes and lizards are shoved into corners on the floor; and bottles of ginseng root preserved in clear yellow liquid decorate the counters. Like the mandrake in pre-Renaissance Europe, ginseng's potency is measured by how much the tuber resembles a person. In Chinese medicine, people become sick because of an internal imbalance between *yin* and *yang*, rather than from external forces like viruses and bacteria.

Traditional Chinese medicine departs philosophically from Western medicine in its emphasis on prevention as opposed to treatment. Even though modern medicine in China is a mixture of traditional and Western practices, pharmacies are busy places; in addition to curative agents, most people take tonics or herbal supplements (something that is today not unknown in the West) to keep their *qi* (the vital life force) strong and yin and yang in balance. In the apothecaries, ingredients are scrutinized, carefully measured, sometimes ground in mortars, and dispensed in small paper envelopes. It takes years of practice to become a qualified traditional doctor or pharmacist. The remedies do work.

Take Mick, for instance. Only later on the fateful day at the Summer Palace did he realize that a chewed straw had run into his mouth with the water. The bottle had been recycled, and Mick was pissed. Perhaps the water was bad, or maybe it was something else. Nevertheless, Mick's health deteriorated quickly. Nine days of regular ingestion of a powerful mixture of Cipro and Pepto-Bismol wasn't helping. So we called on Gao Ke-Qin, who brought some traditional medicine to equilibrate Mick's yin and yang. Gao didn't know what was in it; the standard dosage was one

Smoking of opium and opium laced tobacco has a long and unfortunate history in Asia. Perhaps the only good thing to come out of it is artifacts, like this beautiful pipe, of the rich material culture associated with the addiction.

pill—Gao said take two, because Mick wasn't Chinese and "probably needed more." Shortly thereafter, Mick awoke and was spewing from both ends. The next morning, Mick was still on the bathroom floor, but he was cured. Mick claimed Gao had poisoned him. Later, Gao admitted it.

My most recent encounter with Chinese medicine was during a trans-Pacific flight home. Before the flight, Gao Ke-Qin handed me a bottle of the finest "Bose's Herb Drug Manufactory." He had bought it the day before from a guy in sunglasses behind a subleased store counter. The small vial contained eight tiny yellow *Jin Bu Huan* (Won't trade Gold) anodyne tablets. The label (in Chinese and English) stated: "Action: This medicine is good for anodyne, sedative, spasm, and hypnotic. It is particularly a good remedy for the patient suffering from insomnia due to pain." Just the thing for enduring a cheap trans-Pacific flight, right?

During the layover in Tokyo, before the crux of the journey, I tossed down half the recommended dosage of two microdots with a large glass of Sapporo. (I was already exhausted, so I was cautious.) Going to my seat, things were already getting weird. As advertised, *Jin Bu Huan* is a great sleep-inducer. I don't remember much, except being awakened 14 hours later by the cabin attendant and airplane cleaners at Kennedy Airport in New York, as the last few stragglers collected their belongings. I tried to move, only to feel as if my brain, transmogrified into diced Jell-O salad, would break through my forehead. I felt bad. Not since the morning after a Led Zeppelin concert 30 years ago, where I was the victim of a bad-assed PCP-laced joint, had I felt so bad.

I did learn a couple of lessons. First, don't trust Gao. Second, buy your Chinese medicinal ingredients raw and unprocessed, and have them combined by a *feng shui* master in front of your eyes, rather than prepackaged and created by the Asian equivalent of a meth lab. This experience hasn't scared me off. I still start my mornings with a heavy dose of gingko and ginseng and swear by the curative powers of some of the concoctions that I have sampled.

The bones (*longgu*) and teeth (*longchi*) of dragons are traditional ingredients in Chinese medicine. Their use is recorded as early as the Han dynasty—around 100 CE. As a treatment, *longgu* is powdered and mixed with other ingredients to alleviate extreme dysentery, extirpate ghosts and demonic influences, and cure coughs. *Longchi*, much more expensive due to the comparative rarity of dragon teeth, is used as a tonic for prolonging life and creating a portal to communicate with the *shen ming*, an individual form of intense consciousness. Tactically, it can be given to epileptics and others to alleviate seizures.

Dragon bones and teeth continue to be mined and sold in China. The boxes, drawers, and bottles I have examined have contained a mixture of fossil elephant, rhino, giraffe, and horse bones and teeth. Even some human and primate teeth have been found in this way. Most of these fossils are from the rich fossil beds of Shanxi in Central China and are the remains of animals that lived only about 25 million years ago, long after the nonavian dinosaurs became extinct.

Dinosaur museums are popular all over the world. Crystal chandeliers and drapes form lights and backdrops for these mounted specimens in the Lufeng Museum. This museum is typical of regional museum in China and houses important specimens unearthed in the surrounding countryside.

Dragon images are found throughout China and there is a rich mythology associated with them. Traditionally there are five types: Imperial guardians, the first dragon, earthly dragons, those who control weather, and guardians of treasure. Dragons are capable of changing size immediately and can fly without wings.

In *The First Fossil Hunters*, Adrienne Mayor suggests that dragon mythology in China may have been inspired by the discovery of large fossil bones, some of which were undoubtedly those of dinosaurs. It is clear that dinosaurs were known as extinct reptiles in the Occidental world long before they were recognized as such in China. Furthermore, in Chinese culture, dragons do not have exclusively reptilian characteristics, and are thought of as composites of nine creatures: camel head, deer horns, carp scales, eagle claws, devil eyes, rooster chest, tiger paws, oxen ears, and a long snake neck.

The name "dinosaur" was coined by Richard Owen, the Victorian anatomist and rival of Charles Darwin, in 1842 from Latin words that roughly translate "terrible lizard." The Chinese words for these animals, *kong long* (terrifying dragon), mimic the Western etymology. Both these designations also share an overt recognition that the creatures being described are big, mean, frightening, and ugly. The first dinosaurs to be found were just that, and our contemporary impressions about dinosaurs are little changed, especially the impression that they were all big. But is this true?

The answer is an emphatic no. The pervasive conception of leviathan dinosaurs is rooted in how fossils are collected, how they are displayed in museums, the physics of fossilization, and what kind of environments are likely to produce fossils. The first two issues (collection and display) are interrelated. In the early days of dinosaur collecting, the aim of most dinosaur collectors (and museum administrators) was to find spectacularly large animals with the objective of filling vast museum halls with giant beasts. From the beginning, the public loved dinosaurs, and fossil skeletons attracted scores of visitors to blossoming natural history museums. Bones of large animals were also relatively easy to find and could be collected using brute force methods, including dynamite. Often, the huge, durable bones were collected piecemeal, sacked in bags, and reassembled back at the museum.

But other factors have skewed the proportion of large to small fossil bones, and thus the proportion of large animals to small ones that are collected and eventually exhibited in museums. Physics dictates that large bones are more amenable to fossilization than small ones. Fossilization is a complicated process. In the best-case scenario, organisms are buried alive or entombed shortly after death. This doesn't guarantee fossilization, but when the geological environment is conducive to the process, the result is the kind of spectacular fossils found in Liaoning. This rarely happens. In most cases, when an animal dies, its body is exposed to scavengers,

Small animals, like this bat, decompose quickly. That is why they are so rare as fossils. In China the character for bat and good fortune are homonyms, consequently bat images are symbols of good luck and happiness. Maybe not this one though.

fungi, and bacterial agents, plus weathering, including the destructive action of being transported by water. All of this is hard on bones and teeth, but more so on small bones than on large ones.

Large objects have proportionately less surface than small objects. Generally, decomposition and degradation (like the effects of bumping along the bottom of a rocky stream) act on the surface of a bone, so that big bones decompose and degrade at a slower rate than small ones. This is compounded by the fact that small bones are often more fragile than large ones, again creating a bias toward the remains of large animals being preserved and eventually fossilized.

Finally, there is the factor of where animals live. On Earth today, the greatest diversity of terrestrial animals is found in tropical rainforests. Most of the animals that inhabit these ecosystems are small. Rats and bats, both in number and diversity, far outnumber large animals in such environments. By comparison, in savannah and tundra ecosystems, with their vast herds of hoofed animals, the proportion of large to small animals is more even, and there is less diversity of organisms.

Because the distribution of large rainforests is controlled primarily by latitude, which influences rainfall and wind, it is reasonable to suppose that tropical ecosystems existed in the past. Hot, humid, wet environments are, however, terrible places for fossilization to occur. Bacteria; arthropod, reptile, avian, and mammal scavengers; and fungi all contribute to why Colonel Kurtz's jungle outpost in *Heart of Darkness*, and *Apocalypse Now*, smelled of death, and why tropical organisms disintegrate almost immediately after extirpation, leaving nothing to fossilize. Mental images of the desert usually include a cow or camel skeleton, because more arid environments allow carcasses to desiccate, and the remains are exposed to significantly fewer agents of decomposition. In these environments, bones can lie on the surface for years, and thus they have a much better chance of burial and eventual fossilization.

What this means is that the terrestrial fossil record is skewed and biased toward large animals. That large bones are historically most often collected, that they are most likely to be preserved, and that environments of greatest diversity and numbers

An example of a horse desiccating on the desert floor near one of our particularly scenic campsites in Mongolia's Gobi desert. This one has lasted for years, while the carcasses of small animals disappear over night.

of small animals are rarely sampled show just how biased the fossil record is. This also explains why people generally think dinosaurs were all large.

What about the ratio of large animals to small ones, of large dinosaurs to small ones unearthed in Jehol deposits? Even though they preserve a warm temperate biota, it is one of the few localities in the world where the bias toward large fossils does not hold. Because dead animals were buried so quickly (sometimes by the same volcanic ash that probably killed them), there was little in the way of post-mortem decomposition. This has led to the remarkable situation that lots of small dinosaurs are found in Jehol sediments, and these little animals outnumber the big guys.

The actual fauna of Jehol was undoubtedly much more diverse than is represented in the fossils, yet for reasons of ecological exclusion (habitat preference), or sampling biases like the one just described, remains of very large animals are occasionally found. Furthermore, in ecological communities today, herbivorous animals far outnumber carnivorous ones. But in Jehol, the diversity and abundance of carnivorous theropod dinosaurs is very high, again reflecting some sort of bias.

And then there is the category of very rare animals. While many of the Jehol dinosaurs (animals like *Psittacosaurus* and *Caudipteryx*) are known from multiple specimens, some, like the feathered theropod *Protarchaeopteryx* and the enigmatic armored dinosaur *Liaoningosaurus*, are only known from unique specimens. As the

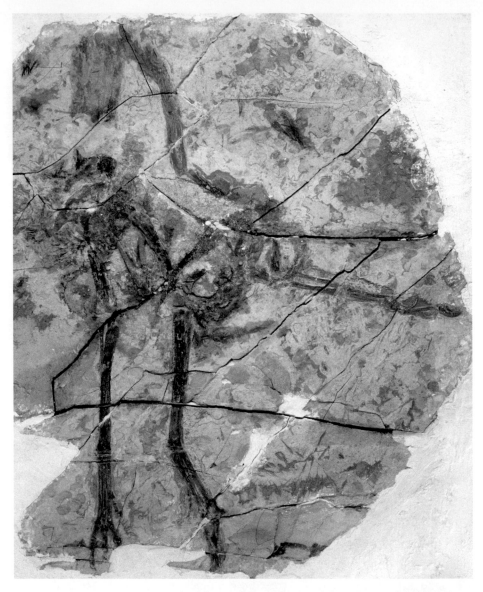

Protarchaeopteryx is an enigmatic and poorly preserved specimen. Nevertheless it was one of the first feathered dinosaurs to definitively preserve feathers of modern aspect.

lower fluvial beds are becoming more densely sampled, especially around the city of Beipiao, more kinds of dinosaurs are appearing as these beds sample a different environment. Here, the remains of some very large animals are becoming more common. These include sauropods and the duck-billed dinosaur *Shuangmiaosaurus* and more small herbivores like the small ceratopsian or horned dinosaur *Liaoceratops*. Sauropods are the familiar dinosaurs with tiny heads, long necks and tails and elephantine bodies. All of these large animals were herbivorous and are distantly related to modern birds.

The occurrence of primitive types of herbivorous dinosaur groups in the Jehol deposits demonstrates what an exciting time it was in the Early Cretaceous. The later diversification of hadrosaurs and ceratopsians into the dominant herbivorous animals of the late Mesozoic Era is correlated with dramatic changes in plant communities. The most obvious is the origin and diversification of angiosperms. It is unsurprising, then, that we find these very primitive herbivores occurring in concert with the evolution of early flowering plants and associated insects. This demonstrates once again that it is not just single animal lineages that evolve, it is entire communities that change together over time.

Xu Xing is one of the young paleontologists working on the dinosaurs of Liaoning. My first acquaintance with Xu demonstrates how much China is changing. During the early 1960s, China suffered through what many believe was the largest famine in human history. This was quickly followed by "The Great Proletarian Cultural Revolution," which wasn't really over until the arrest of the Gang of Four in 1976, and left the country in turmoil. Intellectuals and scientists were among those who suffered most severely during this attempt to reform Chinese society, and it took even longer for the Chinese academic system to recover. The newest generation of Chinese scientists was born well after these events. This, coupled with the globalization of China's economy and the internationalism of its young academics, has created a significant generation gap among its scientists.

In the early 1990s, my Chinese colleagues, even the ones my age, such as Gao Ke-Qin and Meng Jin, had been through the hard times. They were generally soft-spoken and not publicly critical of the work of others, especially of foreigners. Meng Jin is now a curator in my division at the American Museum and has two American children, but he still holds firmly to his Chinese roots; Gao Ke-Qin even more so, as he is often mocked by Chinese old-timers as being the most Chinese man in the world. Much has to do with his encyclopedic knowledge of *kung fu* movies and novels, his fondness for traditional foods like dog, his casual demeanor, and his sense of humor, where all jokes seem to have some moral, whether or not they are funny or even intelligible. Like the traditional Chinese males of the diasporas who founded successful Chinese communities across the planet, Gao adapts, can be separated from his family for long periods, and has the ability to live anywhere. He spent much of his time in New York sleeping in a kitchen of a rent-controlled apartment illegally turned into a boarding house for foreign students. But things change.

The exceptionally talented Xu Xing holding a reconstruction of the head of the feathered tyrannosaur *Dilong paradoxus.*

It was pretty weird then, when almost before I had finished a lecture at the Institute of Vertebrate Paleontology and Paleoanthropology, a hand shot up from the back row. It was the hand of Xu Xing, who at the time was barely more than a teenager. Usually in these situations, the youngest students defer to more senior scientists, only rarely ask questions, and are never critical. This was something different, and before I could even call on him, Xu was arguing with me about my methods and evidence. He became so adamant in his challenge that some of the more senior students in the audience tried to restrain him physically and finally got him to shut up, in a soft Chinese sort of way.

This spirit and his hard work have made Xu Xing an international figure in dinosaur paleontology. Already he has described more of Liaoning's dinosaurs than anyone else. Many of them are rare small animals unique to these deposits.

These small dinosaurs include not only a diversity of types, but, like some of the animals and plants I mentioned earlier, are also very primitive members of groups found elsewhere in the world. This reinforces Zhexi Luo's ideas about this area of Northeastern Asia acting as an ancient refuge, a sort of "Lost World" in Early Cretaceous times, where animal kinds that were extinct elsewhere on the planet continued to survive.

Refugia and island continents are usually home to some pretty weird organisms. In chapter 2, I referred to the lemurs of Madagascar and Darwin's finches of the Galapagos. Because of my background in zoology, there is not much that surprises me in the animal or vegetable world. It is our own species that amazes me. One of my strangest human encounters yet has to do with a fashion craze that swept through Beijing about seven years ago.

In some of the more traditional areas of Beijing, just northwest of the Bei Hai Park, much of the Ming Dynasty city is still intact. The interiors of what were once stately mansions have been subdivided and remodeled repeatedly. Many houses have no running water, and toilets are of the public,

Gao Ke-Qin enjoying one of Cuba's best, while contemplating the dinosaurs of Jehol, or more likely, who is the hottest *kung fu* actress in the world.

communal kind. It is not, however, a slum, and the streets are clean, fanatically swept daily by older men and women in blue shirts and straw hats, wielding brooms. Recently, the area has metamorphosed into a hipster zone, where trendy clubs abound.

On any day that is not brutally cold or rainy, elderly Beijingers sit on low stools and tend their grandchildren, read the paper, play *xiang qi* (Chinese chess), or gossip, drink tea, and smoke. Makai, a connoisseur of Chinese cultural history, often leads us on walking tours, pointing out where Ming walls were reconstructed in the Qing, deciphering the characters on ghost screens, and alerting us to the subtle rolling consonants of old Beijing accents. In the early morning, men congregate with their caged birds to let them sing. Everyone exercises, and in the springtime cricket-singing competitions are waged. Occasionally, an ancient woman with bound feet painfully hobbles by, supported by canes. The elderly residents dress conservatively. Old blue Mao suits are common, though the color palate of most clothes is gray, matching the stone and masonry houses. Walking through the neighborhood is to be transported back to old China and old Beijing.

Imagine my surprise when, rounding a corner, Mick almost bumped into a tall young guy, dressed only in brilliant scarlet pajamas, buying a pack of smokes. His hair was all over the place, his sunglasses were pitched crookedly on his flat nose, and he was wearing way cool faux leopard bedroom slippers. His major accessory was a woman draped all over him, with hair past the hem of her microskirt, balancing on 4-inch heels. "Hottie," Mick pronounced. "Way Hot" I echoed, my eyes still glued to her. This guy really had his game going. It was about 1 in the afternoon, and he had just fallen out of bed after a very rough night and wanted to advertise it. We weren't the only ones staring, as distracted bicyclers narrowly missed gawking pedestrians. Mick pronounced him to be the "Pajama Guy." Surely a once-in-a-lifetime sighting.

Or maybe not. As we walked toward the Forbidden City, we turned another corner, and there was another Pajama Guy, dressed in blue silk, with two girls. Over the next few weeks there were several sightings. The style was so hot that we were momentarily inspired to try to pull it off at home in New York. But style in Beijing changes as fast as it does in New York south of Houston Street. As suddenly as they had appeared, they vanished. That was the spring of the Pajama Guys. Neither Mick nor I have seen them since.

The Pajama Guys were as unexpected and ephemeral as dinosaur fossils can be. In March 2001 I walked into Xu Xing's office. I sat down next to his cluttered desk. He showed me a few fossils and we talked about his upcoming Ph.D. defense, of

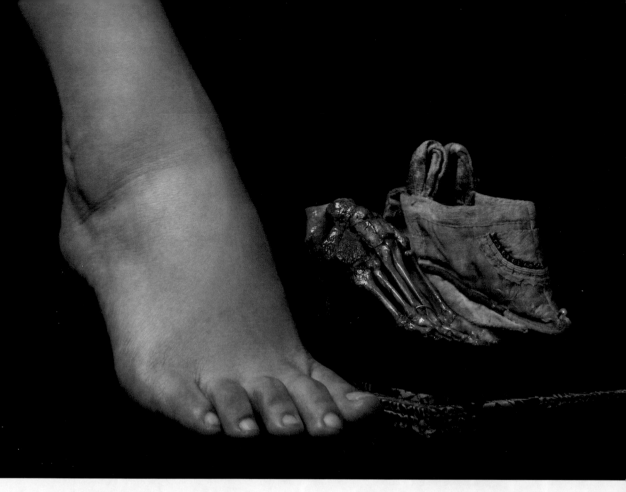

A skeleton of a bound foot next to a tiny shoe with a more or less modern sized foot for scale. This barbaric practice was made illegal in 1911, but persisted through the 20's.

which I was an examiner. After a while, he casually opened a drawer in his desk. In the drawer was a small skull about 15 centimeters long. It did not look like a dinosaur. Unlike mammals, dinosaurs usually have what are called homodont teeth, teeth that are pretty much similar throughout the tooth row. Furthermore, what are called premaxillary teeth, those that are at the tip of the snout, are usually very small, or in many cases not even present. Xu's dinosaur skull looked like Bugs Bunny. It had gigantic choppers at the front of its jaw and only small peg-like teeth farther back, separated by large spaces.

My first inclination was that this was yet another fake, that the teeth of a giant fossil rodent from the Tertiary Epoch had been glued into the head of a much more ancient dinosaur. Even though it looked real under a microscope, I was unconvinced. It was too weird, and I didn't want Xu or anyone else I like getting burned *Archaeoraptor*-style. When Mick and I spotted the first Pajama Guy, we didn't know if he was real or a joke. Just like the reality of Pajama Guys, replication is an

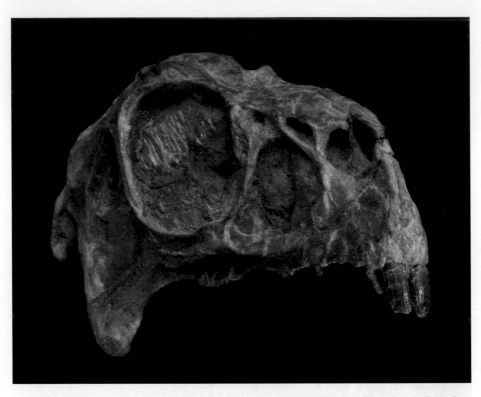

The bizarre *Incisivosaurus*, a dinosaur related to carnivores but with unusual bunny-like teeth at the front of its mouth.

important part of science. I became convinced that what would become *Incisivosaurus* was real, only after Xu reached into his desk and pulled out another specimen.

Even more unusual than the Pajama Guys are animals like the ornithomimid (ostrich dinosaur) *Shenzhousaurus*. This specimen, like the Pajama Guys, is the real deal and is very different from anything that has ever been found. When I saw the specimen in Beijing, it was in cross-section in two meter-long boulders, and I made a preliminary, but incorrect, identification of it as a therizinosaur. Therizinosaurs are rare and peculiar animals. Picture a small head, a fat body with a short tail, and exceedingly long front limbs tipped by three very long claws on each hand, walking on two stout legs. Think of this in an animal covered with feather-like filaments. A small one could look a normal-sized person in the eye. In modern terms, this animal would look like a guy with a tiny head on a long neck with a large beer belly in a chicken suit sporting Edward Scissorhands front limbs. Because of the possible importance of the specimen, Ji Qiang allowed us to ship it to New York for further study.

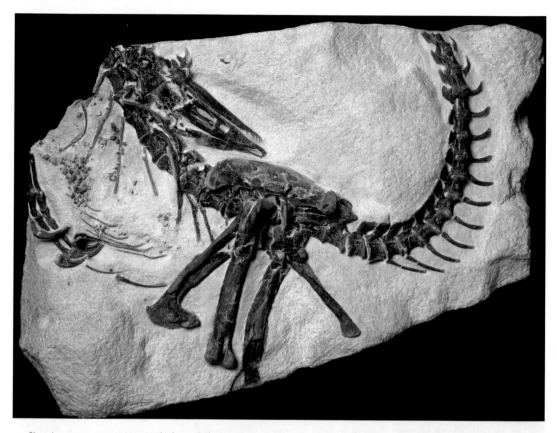

Shenzhousaurus, a primitive toothed ostrich dinosaur preserved on a single large block. This animal lies in a death pose with the head pulled back over the body. This is due contraction of the strong nuchal ligament in the neck. Identical poses occur in modern birds. Check out that dead sea-gull on your next trip to the beach.

Because the specimen was basically sliced down the middle, it was very difficult to figure out what was going on with it. I decided on a strategy that would expose the animal in bas-relief. To do this, we glued the two boulders together, and Brian Roach and Marilyn Fox, technicians who specialize in removing sediments from delicate fossils, began chipping away at the outside, essentially destroying one entire rock, a flake at a time. They revealed a spectacularly preserved ostrich dinosaur, with its head snapped back over its body in what paleontologists call a death pose. Such death poses result from ligament contraction along the top of the neck. In life, these ligaments support heavy skulls and are counteracted by strong muscles of the throat. When the throat muscles relax and the ligament retracts, the head characteristically snaps back. A very long neck often accentuates this pose. Such death poses are common in all dinosaurs, even in living ones. On your next trip to the beach

The small pebbles found in the chest of the *Shenzhousaurus* specimen are gastroliths. Gastroliths are stomach stones that may have been used for grinding food like the stones in a bird's crop.

check out the first dead seagull or pelican you come across; in death, it looks pretty much like the fossil of *Shenzhousaurus.*

Ostrich dinosaurs, which resemble ostriches, are common in rocks that are about 60 million years younger than the Liaoning sediments, especially in Asia and North America. They are bipedal, long-legged, and long-necked, and they have small toothless heads. What was so unusual about *Shenzhousaurus* was that it had teeth. It didn't have a lot of teeth—it was no *T. rex*, nor were the teeth very big. Nevertheless, the presence of teeth in a group that is generally composed of toothless forms is one indication that we had found a very primitive animal. As I have stated throughout this book, primitive animals are the ones we look for, and data from these primitive and transitional forms are what make paleontology so important in deciphering life's genealogy.

Two other toothed ostrich dinosaurs have been found. Both are from slightly younger rocks, one (*Harpymimus*) from Mongolia and the other (*Pelecanimimus*) from Spain. *Pelecanimimus* is a peculiar animal; instead of a few teeth it had hundreds. These small, needle-like teeth are probably flexible. Some soft tissue is also

A pencil reconstruction of how *Shenzhousaurus* may have appeared.

preserved, including a pouch, called a gular fold, under its jaws, just like a pelican. Hence the name. No doubt, this animal was a specialized filter feeder.

Harpymimus is poorly preserved, but we know it had teeth on the lower jaw. This is also true of *Shenzhousaurus*, where only a few teeth lie equally spaced along the lower jaw element. Surprisingly, there are no teeth in the upper jaw. Obviously, *Shenzhousaurus* was a highly specialized feeder, but this is a case where comparison with living animals fails to give us hints about how *Shenzhousaurus* fed, or how it selected its prey. The only extant animals with this tooth configuration are sperm whales, and it is highly unlikely that *Shenzhousaurus'* diet included *Architeuthus*, the giant squid, an animal that lives in oceans so deep that live specimens have never been carefully studied.

Some other things about the *Shenzhousaurus* skeleton do suggest that it was a plant-eater. First, the peg-like teeth would be of little help in subduing prey, the claws on the hands are not fiercely recurved as in carnivorous animals and those dinosaurs that are thought to be predatory. It also has a pile of stones in its abdominal cavity. Such occurrences of stomach stones, called gastroliths, are common in many dinosaur species. Other more advanced toothless ostrich dinosaur specimens preserve the stones. Gastroliths are thought to have acted in conjunction with the stomach muscles, like the stones in a bird's gizzard, as a gastric mill; they were basically a grinding machine for tough, hard-to-digest material.

While this may suggest that *Shenzhousaurus* was an herbivore, things do not always work out so cleanly. Gastroliths are also found in a variety of aquatic reptiles, such as the extinct marine reptile group Plesiosauria and living crocodiles. It has been proposed that in these aquatic reptiles they are used for ballast; however, gastroliths are also found in seals and sea lions, the carnivorous terrestrial theropod dinosaur *Allosaurus*, and lots of different animals. Except for the case of some birds, where they have been directly observed, we have no idea what gastroliths in dinosaurs were used for, so they can't be a direct indication of an herbivorous diet.

Three-dimensional fossils like the *Shenzhousaurus* and *Incisivosaurus* from the lower fluvial Liaoning beds are appearing at a greater frequency. Among these are large concentrations of *Psittacosaurus* babies. Psittacosaurs are called parrot dinosaurs because of their toothless beaks. They were small, pig-like animals whose remains have been found in similarly aged beds throughout Asia. The psittacosaur babies, because they represent several different size classes, suggest young psittacosaurs stayed together as a flock for long periods of time after hatching and undoubtedly needed parental attention. Such social behavior is not present to this degree in more primitive reptiles, but it is in birds.

These animals are some of the most commonly collected dinosaurs from the Liaoning beds and in similar aged beds throughout Asia. Hundreds of psittacosaur specimens have been smuggled out of China, and they regularly appear for sale in high-end curio stores or at rock and gem shows. Most of these specimens are heavily restored chimeras, composed of five or six individual animals pasted together with a lot of plaster and what looks like fiberglass car body putty. I receive at least one solicitation a week for purchase of a specimen, from both inside China and out.

A remarkable specimen of the Jehol plant-eater *Psittacosaurus*. Thousands of specimens have been found, but we are only now getting a clear picture of what the animal looked like when it was alive.

One psittacosaur specimen, mentioned briefly earlier, has even been found with large, porcupine-like spines sticking out from its tail. This is a very controversial specimen, for many reasons. The entire animal is flattened on a single slab, and a large amount of soft tissue is preserved, showing that most of its body had a scaly texture composed of small, plate-like scales surrounding roundels and quadrangular tubercles. Such a skin texture is known from a wide variety of dinosaur specimens collected around the world.

A second sort of soft-part preservation is a horny sheath that covered the horns on the side of the skull. This is much larger than the underlying bone and would have given *Psittacosaurus* a fearsome warthog-like appearance when alive.

It is the third type of soft tissue that is most interesting. It consists of long (up to 16 centimeters) bristles that line the top of the tail. The bristles are not branched. Structures like this are so weird that they strain incredulity. It was even suspected that they may have been part of a fossil plant crushed on top of the *Psittacosaurus* skeleton. What they were used for is highly speculative. Just as archeologists when confronted with an object that they can not identify often suggest it had a ritual or ceremonial use, paleontologists chalk it up to display. Having examined the specimen firsthand, I can attest that the structures are real and neither forgery nor plant.

The spiky psittacosaur specimen has achieved some notoriety for other disquieting reasons. It was smuggled out of China and offered for sale at the Tucson Gem and Mineral Show in 1998. The illegal traffic of Chinese fossils from Liaoning and other sites has been intensely reported by Rex Dalton, a *Nature* magazine correspondent, in a series of articles.

My first introduction to Rex was at the Society of Avian Paleontology meeting in Beijing in 2000. Contrary to American stereotypes that Chinese are basically short, the people of China (especially northern China) are a tall race. Still, at well over 6 feet, with a booming voice, Rex stands out in any crowd. At the end of one day's sessions, we headed out to one of Beijing's night markets, where cheap foods of all kinds are served from the backs of bicycles. Night markets operate for four or five hours each evening and traditionally cater to laborers and the like. In recent years, many markets in the larger cities have become upscale and are packed every evening with young locals, families, and tourists.

At this night market, I learned that Rex will eat anything and has a hollow leg. Rex scarfed down fried scorpions, embryonic quail, silkworm larvae, and a giant cuttlefish on a stick. He even ate *cho dofu*. Although it was one of the favorite foods of Deng Xiao Ping, *Cho dofu* tops my list as the foulest food on the planet. Most Chinese have an aversion to cheese. They especially hate pungent, strong-tasting cheeses like blue cheese or ripe camembert. I have never understood how they can stomach *cho dofu*. It is just like cheese, but soy-based; it is tofu that has been allowed to rot under controlled circumstances to form a black pillow of rancid, musty goo. Rex ate it.

That night he was an eating and drinking machine. He saw a chicken foot, he had to have it; he saw a grasshopper, down it went. Then he saw the bowls of Hunan-style noodles. He had to have one of them. In one hand, he had a beer and chopsticks as he was handed a large ceramic bowl of noodles soaking in chili sauce, so red that it was like peering into the crater of a volcano. The street was packed. What Rex didn't realize was that the boiling noodles were actually significantly cooler than the bowl itself, which approached the temperature of the solar surface. After a couple of seconds of painful, but funny, dancing around, he dropped the bowl. It didn't drop straight down, and hot noodle broth spilled all over Rex before the bowl exploded on the ground. It was like an Old Faithful of fiery red oil and noodles going everywhere. People scattered and were angry. A woman in a soiled white silk dress had to be restrained.

Rex is working on paying off his karmic debt to the Chinese by doing a great job of bringing the illegal fossil trade to light. The psittacosaur specimen is a good example. Rex documented its underworld journey from China to California; from there to Tucson, Ariz.; then to Trieste, Italy; and finally to Frankfurt, Germany. The specimen was purchased by the Senckenberg Natural History Museum, in Frankfurt, for an amount reported to be around $200,000. The sale of this fossil has become a major source of irritation to Chinese scientists—and me. The international scientific community has called for repatriation of the specimen and condemnation of the Senckenberg museum for colluding in the illicit trade of fossils.

The Senckenberg specimen is, however, spectacular, in that it conclusively shows a fringe of spiky structures along the animal's tail. The remainder of the body is covered with typical scales, like those seen on some dinosaur mummies or on the feet of living chickens. Psittacosaurs and their relatives, the horned dinosaurs, are also known for having spikes and frills on their heads. In primitive members of this group, the largest spines are the ones that stick out on the edges of the skull just behind the mouth. On this specimen, it can be seen conclusively that these horns, called the jugal horns after the bone that supports them, were covered with a horny

The night market. Night markets are popular places for light meals and socializing all over China and almost anything that can be consumed is.

American Museum of Natural History preparator Jason Brougham next to his unpainted, ghostly model of *Psittacosaurus*.

sheath in life. This sheath, similar to the sheath of a cow horn, was probably composed of keratin, the stuff of modern horns and hair, feathers, and beaks. These sheaths would have increased the size of the horn at least twofold, giving the psittacosaur an appearance akin to a combination of a pig's body, with the tail of a giant anteater and the head of a horned lizard.

The tail filaments on the Senckenberg psittacosaur are still being studied, as their presence has vast implications for all dinosaurs. They hint at the presence of integumentary structures like fuzz, filaments and feathers on all dinosaurs. From other specimens found at Liaoning and elsewhere, we know that pterosaurs (which are dinosaur relatives) and some dinosaurs had integumentary structures of filaments or feathers. An explanation for this is that this sort of feature was present in the ancestor of all of them and therefore would be expected to occur throughout their diversity.

In an earlier chapter we mentioned the sleeping troodontid. We named this animal *Mei long*, which means sleeping dragon in Mandarin. I refer to it here because the preserved behavior of stereotypical tuck-in sleeping, seems to suggest that it had feathers and that this posture was one that minimized the surface area from which metabolic heat could escape. A look at other less well-preserved specimens suggests that a few other small Chinese troodontids were also preserved in this style, such as *Sinornithoides* from slightly younger rocks of the Ordos Basin in Inner Mongolia. The best story for these guys is that they were warm-blooded feathered animals preserved Pompeii-style. While this difficult to test, it is strong inferential evidence that these animals and their relatives were warm-blooded.

But what about feathers? Most of the fossil dinosaurs unearthed at sites around Liaoning do not have preserved soft tissue. But from what we can determine about the relationships between dinosaurs and birds, and the empirical evidence we have been able to garner from the Liaoning specimens, it appears that many dinosaurs looked a whole lot different from the way we thought just a few years ago.

The three dimensional specimen of *Mei long* from the "Pompeii beds" of the Yixian Formation. Preserved in a characteristic resting position where the head is tucked in between the elbow and body, perhaps this animal was killed by noxious gasses before a quick burial.

CHAPTER 7

Feathers

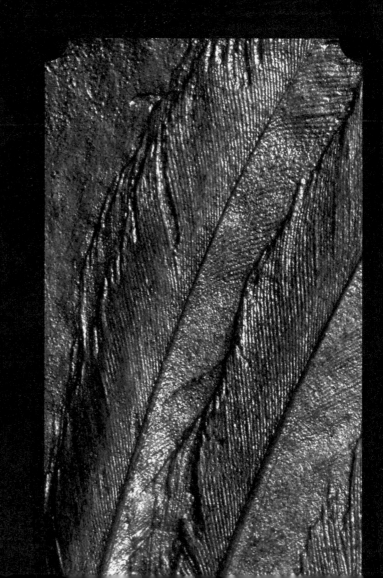

Sonic Hedgehog and Showgirls

The initial Western announcement of the first "feathered dinosaur" by Chen Pei Ji and colleagues in the journal *Nature* in 1998 surprised no players in this game. My colleagues in paleontology in China, the United States, and around the world had been alerted to the existence of the specimens two years before, by Chen himself at the 1996 Society of Vertebrate Paleontology meetings in New York. The *Sinosauropteryx* specimen had already been named (based on the counterslab) by Ji Qiang and Ji Shuan in the *Journal of Chinese Geology* in 1996. Unfortunately, this journal is narrowly distributed, and the article was in Chinese. It was off the radar of the Western press. This fairly primitive theropod dinosaur had a covering of "fuzz" over its entire body. The discovery of several more new kinds of animals that not only had fuzz but fully formed feathers, just like modern birds, quickly followed.

Ji Qiang had e-mailed me from China in the spring of 1999, telling me he had something great, and Mick and I should get over as soon as possible. Mick had gone on ahead, and a week later I was on a plane headed

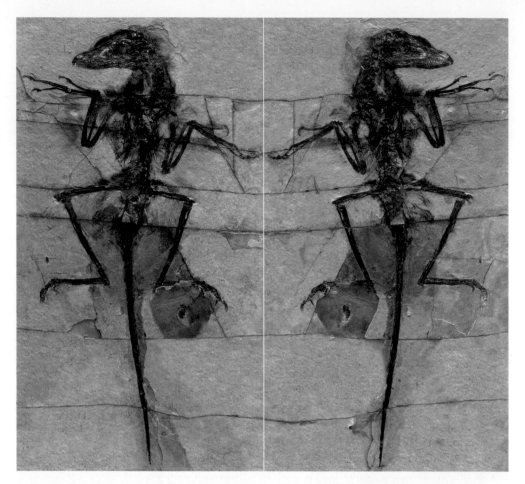

Dave the feathered dinosaur. Spread out in a crucifixion pose, it preserves a feathery covering over its entire body. Many have commented that Dave is one of the finest dinosaur fossils ever found. While it is not easy to determine whether or not Dave is a new kind of animal or a juvenile of the dromaeosaur *Sinornithosaurus*, this specimen has the most complete body covering of any non-avian dinosaur yet discovered.

for Beijing. I arrived at the Xiyuan, a hotel near the Beijing zoo, jetlagged and coming down from the trauma caused by a 24-hour journey in coach. My body wanted to sleep. My brain, however, wouldn't cooperate. In Beijing it was midnight; it was noon, Manhattan time. I crept into the lobby, only to find Mick, Gao, and Ji looking half-crazed and clearly celebratory. The *gu niang* brought me a beer, and Mick, supporting his stubbly chin with one hand, could only slur "You gotta see it, you gotta see it!"

It wasn't clear what he was talking about until we went into the basement of the Geological Museum the next morning. The halogen photo lights went on with the subtlety of a nuclear explosion in my eyeballs. Out of the red and blue dots, the greatest dinosaur specimen I had ever seen came into focus: a small carnivorous dinosaur completely covered with feathers.

We went to a long lunch with Ji Qiang and a few of his colleagues. Although Ji Qiang certainly realized that this new animal was important and that it was not a bird but a feathered dinosaur, the rest of the crew was confused. They had not moved on from the character-based definition that anything with feathers is a bird. I was tired, and the more I tried to explain it, the more confused we got. Such confusion was not an unfamiliar experience for us in China. Our cultural baggage and lack of a common language collide and blow right past each other.

Mick and I retreated to the Xiyuan, a nap, a massage. Digging into the inner recesses of the mini-bar we reflected on the afternoon's conversation. We agreed that the afternoon's conversation could be summed up by a flashback to an old Cheech and Chong routine—one where Cheech Marin attempts very unsuccessfully to communicate with his friend Tommy Chong about the whereabouts of a character named Dave.

In light of his cultural reference point, that evening we decided to give the specimen the colloquial name "Dave." Our work on Dave resulted in a paper in *Nature* in 2001 with the prosaic title, "The distribution of integumentary structures in a feathered dinosaur." Many consider Dave to be the finest dinosaur specimen ever discovered. It is so crisp and reads so well that you can show it to anyone. I have had 5-year-olds in my office, creationists, and even molecular biologists. All of them, without prompting, can identify Dave as a feathered dinosaur.

What makes it so special is that it is splayed out on two slabs of rock, part and counterpart, crucifixion style. Dave is a dromaeosaur, a member of a group of dinosaurs that are closely related to birds. It is one of those frustrating Jehol specimens that are only visible in two dimensions, so I decided to be conservative and not assign it to any known species, nor to erect a new one. The differences between it and another Jehol dromaeosaur, *Sinornithosaurus*, are minimal. Furthermore, Dave is smaller than the *Sinornithosaurus* specimen, and the apparent differences may be explained by allometry, like the differences in relative neck length between the baby and adult *Hyphalosaurus*. The arms are extended out from the body, and the entire torso, head, arms, legs, and tail sprout long feathery filaments.

These filaments can be categorized into three types. The first are simple fibers, like the ones in *Sinosauropteryx*. They occur primarily on the head and tail. Second, sprays of fibers (groups of filaments from a common origin) are preserved on the torso, arm surfaces, and legs. They are especially thick on the hind limbs and shoulders. But most surprising was the presence of hints of feathers of "modern aspect" (identical to those in a modern bird) on the trailing edge of the arm. Their presence

A reconstruction of the big theropod dinosaur *Sinraptor* outside the Institute of Vertebrate Paleontology and Paleoanthropology in Beijing. The building across the street is the Xiyuan Hotel, home of the infamous and defunct Zui Zui Le karaoke lounge.

was confirmed when my colleagues and I described a second dromaeosaur from slightly younger Jehol rocks that preserved the definitive presence of real feathers on a nonavian dinosaur.

This second specimen, which we call Chong, was described in a short *Nature* paper in 2001, simply titled "Dinosaur Feathers." Like Dave, this specimen preserved something scientifically informative (the feathers), but it just was not preserved well enough to call it a new animal. Later, someone else, unfortunately, did name it. Stuff like this is a disservice to paleontology, as creating a new name and type specimen of something so devoid of character information accomplishes little, and in some sense diminishes our ability to use quantitative analyses of genealogic patterns, because there is just so much missing data. James Clark of George Washington University put it best, when he said in disgust, "As much progress would be made by putting all of the shitty-type specimens in one place and blowing them up, as would the collection of a few more new fossils."

Because Mick and I have been at the center of many subsequent discoveries of feathered dinosaurs, there is much to say here, beyond merely describing the animals. These discoveries have changed the way everyone thinks of dinosaurs and have important ramifications for the origin of flight, the origin of warm-bloodedness, and the origin of birds.

The feathered animals from Liaoning are some of the most unusual that anyone could ever imagine. *Caudipteryx* has a long, striped tail plume, a tiny head, and four small teeth at the end of its beak. *Beipiaosaurus* has goofy enormous forelimbs, nasty claws, and strange "feathers" sprouting from the backs of its arms. *Microraptor gui* has feathers extending from its hind limbs, presumably giving it the appearance of four wings. Was it a biplane glider? The list goes on.

Outside China, people were largely unaware of these animals until 1997, when the Academy of Natural Sciences in Philadelphia sent a panel composed of American and European paleontologists to China to view the specimens. On their return, they called a press conference announcing their findings. I felt that it was an incredibly hubristic act on the part of the organizers and participants to have a press conference validating specimens already described by Chinese paleontologists, especially because not a single Chinese scientist was invited. This was intellectual colonialism. At the time, Ji Qiang was visiting me in New York, and when I got wind of the press conference, we jumped on a train to Philadelphia and just showed up.

The press conference didn't go all that well. Members of the panel disagreed about whether the specimens they had seen exhibited feathers. Things really got bad the next day, when the story was published. The April 25, 1997, front page of *The New York Times* featured a story in the coveted upper left-hand column headlined, "In China, a Spectacular Trove of Dinosaur Fossils Is Found." The article, written by Malcolm Browne, reported, "An international team of paleontologists announced today that a fabulous trove of dinosaur fossils had been discovered in a remote region of Northeast China."

Caudipteryx zoui is this unusual dinosaur that was the first to show unambiguous evidence for feathers of modern aspect on a non-avian dinosaur.

Feathers have long been used as ornaments. Here the iridescent feathers of a kingfisher adorn the surface of a Qing dynasty hair pin.

The reference to "an international team" wasn't to Ji Qiang and his Chinese colleagues, who had actually discovered these specimens, but rather to three American paleontologists (John Ostrom of Yale University; Alan Brush, then of the University of Connecticut; and Larry Martin of the University of Kansas) and one German (Peter Wellnhoffer of Munich). Ji was relegated to a single quote at the very end of the article, only because he happened to show up. This was wrong, whether you are Chinese or American, paleontologist or publisher, scientist or regular human being.

All the disagreements, excitement, and hype surrounding the Jehol dinosaurs boils down to one thing: feathers. Feathers are considered by many to be an almost magical adaptation. Their usefulness for birds is apparent, and their beauty has been used to decorate everything from Vegas showgirls to Trobriand islanders. Iridescent blue kingfisher feathers were even one of the dominant materials of Qing costume jewelry and decoration, often highlighting elaborate hairpins and the margins of mirrors.

To many people, feathers define birds. This definition brings us to the source of a major misunderstanding within modern systematics, the process of classification. Before evolutionary theory became accepted in the mid-19th century, the dominant form of classification was typological. In a typological system, fixed key physical characteristics define groups. For instance, characteristics like feathers in birds or scales in fish defined these groups.

Evolutionary thinking changed all this. Instead of fixed characteristics defining groups, the characteristics were understood to evolve into one another. Fins over time could possibly be transformed into arms and wings. The implications for this regarding taxonomy are clear. It is not characteristics that define groups, it is ancestry. Dinosaurs are called dinosaurs because all of them are more closely related to one another than any is to a pterosaur. Or another way of putting it is that dinosaurs all share a common ancestor more recently than any is related to a pterosaur.

This brings us back to the question that Ji Qiang posed to me at our first meeting: "Feathered animals are bird or dinosaur?" There are several answers to this question. The first is that birds are a kind of dinosaur, just as we are a kind of mammal. The second is that the origin of feathers and the origin of birds are decoupled. The third answer is that the definition of birds has no scientific meaning, *sensu stricto*. For one reason or another, as witnessed by the legions of passionate bird-watchers, birds hold an iconic position within human culture. By calling birds a type of reptile, I was seen by some to equate them with those that crawl, a point of view not too popular with the Audubon set. Nor does it sit well with some others. Shortly after I began publishing papers on this issue, I had an objection from a Brooklyn rabbi, who pointed out that if birds were reptiles, chickens and other poultry would not be kosher.

Even within the scientific community, the term "bird" means different things to different people. The presence or absence of feathers is not, however, the deciding factor in resolving these differences, because we know that the Liaoning deposits have yielded fossils of animals with feathers that are much more primitive than our living feathered friends. What is required is a new terminology based on systematics of ancestry rather than typology.

A consensus has been developing to use the term "avian" for all of the descendants of the last common ancestor of the living bird diversity. These are the animals that many of my paleontology colleagues, and most ornithologists, conceptualize as birds: animals that have beaks, short tails, fly (for the most part), are warm-blooded, and take dumps on your car and outside Ji's office. Animals more primitive than avians, those animals that share a common ancestry with *Archaeopteryx* to the exclusion of all other creatures, are called avialans.

The Liaoning fossils conclusively demonstrate that feathers occurred on animals that came earlier than either avians or avialans, and in the formalism of science, we are left with a world without birds. In the pragmatic world, we can equate bird with

Feathers are among the most versatile of objects. They provide many different functions in modern birds as their varied colors, shapes and structures attest.

avian, even though this leaves out a whole lot of creatures that flew, were feathered, and/or brooded their nests, creatures that, if they were around today, you would definitely call birds.

Even though feathers do not a bird make, they are unusual, intricate, and highly structured features. They are composed of keratin (the same as your fingernails). In living birds, they are used for a variety of functions, including display and insulation, camouflage, and as material for nest-building. The typical feather is composed of a hollow shaft; filaments called barbs extend from this shaft. These barbs are organized into the familiar vane of a feather. Anyone who remembers the hot-tub parties of the early 1980s knows that feather vanes, even long ones like peafowl tail feathers, can be put back together again, even after a good drenching. This is due to the presence of minute, hook-shaped structures called barbules. Barbules work like organized Velcro, or a zipper, allowing a feather to be put back together after it becomes disorganized.

Of course, there are basic modifications to this structure. Some feathers, such as down feathers, function primarily as insulation and lack barbules. Other feather types even lack barbs and, consequently, vanes; these are just like small filaments.

The feather covering of modern birds is heterogeneneous as in this crow with different kinds of feathers distributed over different parts of the body.

They are the sorts of feathers found around the beaks of such birds as flycatchers. The vanes of wing feathers are asymmetrical, with a leading edge that is shorter than the trailing edge. This helps give the wing an aerodynamic contour, generating lift and allowing powered flight.

The evolution of so complex a structure has until recently been poorly understood and hotly debated. Feather evolution, especially the incredible complexity of the design and how perfectly suited different feathers seem to their different functions, bothered even Charles Darwin. In a letter to Harvard botanist Asa Gray, Darwin wrote, "Trifling particulars of structure often make me very uncomfortable. The sight of a feather in a peacock's tail, whenever I gaze at it, makes me sick!" His feelings were representative of other Victorians and were echoed by the naturalist Alfred Russell Wallace (who independently proposed the theory that has become known as Darwinism), when near the end of his life he stated, "Evolution can explain a great deal; but the origin of a feather, and its growth, this is beyond our comprehension, certainly beyond the power of accident to achieve."

Progress has been made. Until recently, the general sense was that bird feathers evolved from reptilian scales. Most of the scale-to-feather evolutionary models suggest

Mick Ellison studying Dave in the basement of the National Geological Museum of China.

that the primitive feather evolved, and then this feather diversified into the multifold sorts of feathers that form the plumage of modern birds. Yet this idea, as pointed out by Yale's Richard Prum and now-retired University of Connecticut zoologist Alan Brush, is fundamentally flawed, because each of the modern types of feathers shows peculiarities that render it inadequate as a primitive or ancestral morphology, relative to other feathers. There are also extreme differences between the structure and development of reptilian scales and avian feathers.

Prum went further, to produce a model of feather evolution that does not consider feathers to be directly related to modern reptile scales. Instead, Prum proposes that both feathers and scales develop from the same primordium (undifferentiated cell mass in developing embryos), and that feathers themselves differentiate early in the development of organisms that have them. I know that this is difficult, but language can be seen as an analogy. Modern Japanese and Chinese share many of the same characters. Written Chinese appeared first, with recognizable characters going back to the Shang and Zhou dynasties, about 3,400 years ago. Some of these are inscribed on bones and turtle shells and used as oracle bones for purposes of divination and later in the Zhou dynasty as carved inscriptions on bronze vessels. It was not until the sixth century CE, that Chinese writing was exported to Japan, which at the time lacked a writing system. However, the writing of the Tang dynasty in China is very different from today's writing, and in no sense would we say that Japanese "evolved" from the modern Chinese script. Instead, as with feathers and scales, it and modern Chinese evolved from a precursor to the modern script. In the literal sense, it is historical time.

Prum's model of feather evolution, then, is intimately tied to feather development, which begins as a thickening of the outer layers of the skin, the dermis and epidermis. Proliferation of these cells forms a bud that is followed by the bud lengthening and the appearance of a moat-like cavity (called the follicular cavity) surrounding it. This results in an architecture where the bud is composed of a dermal tissue layer with an epidermal layer on the outside that is separated from another epidermal layer by the follicular cavity. As growth continues, the bud lengthens and the dermal interior disappears. The bud becomes tubular. Finally, further embellishments of the adult feather structure are determined by basic modifications in the structure of the epidermal collar on the skin surface.

What are these modifications? Just in the past few years has the molecular signaling mechanism responsible for modifications in feather structure been identified. As understood, it is the synergistic effects of four proteins—noggin, BMP4 and BMP 2, and the creatively named Sonic Hedgehog. Like mountaineers naming climbing routes, molecular geneticists can coin or apply any name they want for novel new genes and proteins.

These proteins are common throughout animal life and are involved in basic pattern formation of several structures. These proteins interact to form the complete feather. BMP4 is involved with the development of the rachis—the central shaft of the feather—and the annealing of the nascent barbs to it. BMP2 is involved in the three levels of branching required to form the adult feather. Noggin is also involved with rachis formation and causes the barbs to branch. Sonic Hedgehog is a gene that is responsible for segmentation in many areas of the body during development. In feathers, it causes the barbs to appear in intervals along the rachis. It is thought that the interplay between concentrations of noggin and BMPP4 determines how many and how big the feather barbs are.

A combination of the molecular evidence and Prum's model strongly suggests a four-step scenario for feather origin. Stage 1 is the formation of a shaft or single hollow filament, formed from the thickening of dermal and epidermal structures; in stage 2, bunches or sprays, lacking a rachis, develop from a single follicle (probably influenced by activity of BMP2). Sequentially, this is followed by stage 3, barbule formation, and finally, stage 4, the formation of the rachis.

It is an exciting time to be a paleontologist, as so many new sorts of data, like that developed by Prum and Brush from the delicate world of developmental biology, informs and is informed by the fossils. At last, we may be able to test some aspects surrounding the origin of feathers.

CHAPTER 8

戴维

Dave

And the Theropod With Four Fingers

Now that we have an idea of how feathers evolved, we can begin to entertain the more difficult question of why.

The obvious answer to why feathers evolved would seem to be for flight. The first known dinosaur to achieve flight (with the possible exception of the problematic *Microraptor gui*) was *Archaeopteryx*. Although not all paleontologists agree that it had the aerodynamic abilities of modern birds, most would agree that it was at least capable of some flapping flight. Most phylogenetic, or family, trees place feathered dinosaurs in traditional nonavian dinosaurian groups, i.e., with dinosaurs, outside the group that includes *Archaeopteryx*. It is generally thought that feathered nonavian dinosaurs did not fly. For most, it would have been impossible, as their forelimbs were much too short. What this means is that feathers first evolved for some other reason than as an adaptation for flight. The best candidate is that they functioned as a thermal barrier for insulation.

Mick Ellison's pencil reconstruction of a very feathered Dave. While we have fossil evidence for the body covering, the spots and stripes are conjecture. Does this look like your typical dinosaur?

The fossil of the "four fingered theropod" (nicknamed Chong, after Tommy not Ji) turned out only to have three fingers. It was, however, the first evidence of large feathers on the hind limb. These feathers were of modern aspect just like those on today's birds.

What can we say about the distribution of feathers in other dinosaurs? We can make a strong theoretical case that many were feathered, or at least fuzzed, even though no fossil feathers have been found associated with them. The reasoning is that we have found other close dinosaur relatives that do have feathers, so the simplest explanation is that these animals, too, were feathered. It is the same argument used to theorize that Lucy, the archetypical fossil hominid from Afar, Ethiopia, had hair, although no fossil hair has ever been found with any australopithecine specimen. Chimps and people are both hairy. Because the species to which Lucy belongs (*Australopithecus afarensis*) is descended from the same common ancestor as chimps and me, and, going back further, cheetahs and me, or even further, guinea pigs and me, all also hairy, we say she was hairy.

In early 2001, we heard from Ji Qiang that an amazing new specimen of a feathered dinosaur had been found near the town of Chaoyang. The rocks around Chaoyang belong to the Jiufutong Formation, and although they are similar to the Yixian rocks, they are a few million years younger. What made this specimen so intriguing to us was that, in addition to feathers, it was said to have four fingers. Birds today have remnants of three fingers, and the nonavian dinosaurs to which they are related, animals such as troodontids and dromaeosaurs like Dave, have the same number. Only very primitive theropod dinosaurs, animals more distantly related to birds than even *Tyrannosaurus rex*, have four or more fingers. If it were true that a four-fingered animal did have feathers, it would mean that feathers were a very primitive dinosaur feature, and their presence could be pushed even further down toward the base of the dinosaur family tree. This would be a watershed discovery, because it would indicate that nearly all theropod dinosaurs had feathers.

After arriving in Beijing, we boarded the night train to Shenyang. Chinese trains are divided into several classes, and the only available "seats" were what is called "soft bed." Soft bed amounts to rows of bunks three high, with six sharing a common side aisle. More people are traveling in China than ever before, and all of these trains are packed. To get on one you need to battle your way through the narrow passageway

Trains are one of the primary modes of intranational travel in China.

and traverse mountain ranges of luggage. Everyone is either bringing home souvenirs from a visit to the capital or taking gifts to relatives in the hinterlands. Especially dangerous are trains running around the lunar New Year, when they have been known to blow up because of the explosive presence of contraband fireworks. Our navigation of the narrow path between bunk rows is complicated by bulky photographic gear, scientific equipment, and computers. I lead the way, carrying a formidable-looking 24-kilogram microscope stand and a Yanjing beer. People see me coming and scatter.

Top bunks are choice, but they come at the price of very hot dead air. The trains are "no smoking," so the vestibules between cars become the smoking lounges. Every time the door opens, smoke gets sucked into the cars and equilibrates, so vestibule and coach are equally smoggy. I won't even talk about the bathrooms. The top bunks are so close to the ceiling that it requires Houdini-like contortion to get into bed, and you can forget about rolling over. They say there are middle bunks on these trains, but they are seldom seen. Instead, the lower bunk is used as a combination of bench, card table, bar, and snack counter. Few people sleep, and all consume whatever is brought down the aisle or sold through windows at station stops; many drink. Somehow, Ji Qiang is impervious to it all, and is able to get a good night's sleep (on a top bunk) without taking his tie or sweater off, looking refreshed in the morning. I usually slouch over in the corner of a lower berth and get awakened two or three times by someone using my real estate as a step.

When the morning came, we were in Shenyang. We took a cab to the hotel that would house our makeshift laboratory and photo studio for the next several days.

Before we even looked at the specimen, we were worried about our voltage-sucking bulbs. We were on the fourth floor of the hotel, and one look at the wiring didn't give us a lot of confidence about its infrastructure. We could only imagine the best-case scenario of plugging our big halogen lights in and the building going dark. A darker view would be spontaneous combustion of the polyester carpeting. The windows had bars, and furniture blocked our fire exit—we had moved it to make space in which to operate. The manager was also apprehensive. Before we plugged in, he marshaled meters of extension cord to distribute the power over several circuits. With a crackle, the lights snapped on, it was show time, and we were ready to get famous.

I have been asked what it is like to discover new fossils, or to be shown something from deep time that hasn't been recognized before. In my business, as in much of science, so called "Eureka" moments are few and far between. The best discoveries are those that come out of long planning and a lot of work. They are almost always

The deceiving hand of the four fingered dinosaur. The arrow points to a feather impression that could be mistaken for a fourth finger.

expected. Sometimes, such as when I saw Dave or when I discovered the embryo of the oviraptorid in the Gobi Desert, you know you have something great. But you still have to do the research, get the papers published, evaluate your evidence on its merits, and combine it with whatever else has been done. Usually, by the time a paper comes out and someone wants to know how you feel, a) you are sick of it, b) it is out of date, and c) you have moved on to something else. Besides, when someone describes something to you that sounds too good to be true, it usually is. The fossil you are shown never quite matches up to the mental image that you create for it.

It is good for my blood pressure that I have learned to be pessimistic. The four-fingered theropod was no such animal. It was evident that a section of a feather lay parallel to the three finger bones, appearing enough like a finger to dupe some of those who first looked at the specimen. The specimen was preserved on two slabs with all of the characteristics of a dromaeosaur, including elongate articulations of the tail bones and an enlarged second toe, but beyond that it was not a great specimen. Like so many of the Jehol specimens, most of the bones were in lengthwise section. It is like if you wanted to study the crust of French bread and all you had to work with was a loaf cut down the middle with each half embedded in concrete exposing only the white fluffy part. Much of the detail had been destroyed by an attempt at molding it—the entire surface had been coated with a gooey mess of petroleum jelly (much of which clogged the cracks) to act as a mold release.

But all was not lost. Like Dave, it had a fluffy body covering. Unlike Dave, however, the "feathers" were distinct, and in a few patches had the unique structure of vanes formed around a central shaft that is the same as the architecture of modern bird feathers. In Dave, the presence of feathers of "modern aspect" was suggestive; here it was conclusive.

The specimen was hard to interpret. Its arms were folded back across the body, and feathers and skin tissue were spread everywhere. I was trying to decipher it, trying to figure out which bones went where, and with what feathers. I studied one slab, feverishly typing notes, while Mick was photographing the other. Some of the details just did not make sense. The place where the feathers were the longest and best-preserved was a patch, below the tail next to the hindlimb. If these were the evolutionary precursor to the primary flight feathers of modern birds, all of my experience told me that they should be on the trailing edge of the arm. Right? Was this a patch of feathers that had torn off the arm and been dislocated to the hind limb?

After considerable consternation, and after laying the two slabs next to one another, Mick convinced me otherwise. If you traced out every single one of them, long feathers were present both on the fore and hind limb. Before I published this, it took a lot to convince Gao Ke-Qin and Ji Qiang that it was true. A few months later, in March 2002, our paper appeared in *Nature*, describing both the unusual distribution of feathers and the fact that the feathers showed characteristics of having the microscopic barbs that aid in feather self-organization—the zipping and unzipping of the feathers themselves.

While most paleontologists didn't question the fact that these were true feathers in a definitively nonavian dinosaur (except for the usual recalcitrants you will read more about later), several questioned the distribution of feathers on the body. Steve Czerkas even suggested that the "feathers are not from the hind limbs as Norell suggests" and that my "misinterpretation of the primary wing feathers as being from the hind legs stems directly to seeing what one believes and wants to see."

A pensive, thoughtful and quiet Bao, thinking how great that last cup of baijiu tasted.

顾氏小盗龙
Microraptor gui
时代：早白垩世
产地：辽宁朝阳

The remarkable thing about this feathered dinosaur, *Microraptor gui*, is that, like Chong, there are extensive feathers on the hind limb.

This is laughable, because how could I "see what one believes" when I had no idea what I was looking at. Our discovery of the wing-like hind-limb structure was something I had never seen and would never have imagined.

It's like hearing slang in a foreign language. A few months ago, we were hurtling down the highway at well over 150 kilometers per hour, Metallica blasting through the blown speakers of an old Japanese SUV, toward Beipiao. My friend Bao was at the wheel, and Mick and I were chattering away. This was dangerous stuff, because all kinds of obstacles lie on the road. There are farm animals, people, military vehicles, and large trucks of live caged dogs, destined for the restaurants of Korea. In the middle of all this, Bao turns to me, cigarette in hand and says, "Mark, what means okie-dokey?" It scared the hell out of me, because Bao is one of those who like to look you in the eye when he talks. I wanted him to look at the road, so I answered quickly.

The point is that the phrase was something brand new, a slang phrase he had never heard, and he didn't know what to make of it. Just as I didn't know what to make of the large hind-limb feathers on Chong, it is preposterous to suggest that this is something I wanted to see.

Shortly after I submitted the paper on Chong to *Nature*, I was at the Institute of Vertebrate Paleontology and Paleoanthropology in Beijing, and Xu Xing showed me some new specimens in his office. These were from a similar, closely related animal from the older Yixian Formation. These specimens were better preserved, and it was impossible to question the distribution of feathers on these animals. Better yet, long feathers of modern aspect were also present on the hind legs.

When the paper came out describing this specimen as *Microraptor gui*, it hit the press worldwide and was hailed as the four-winged dinosaur. One of the conclusions made by the authors, Xu Xing, Zhou Zhonghe, and others, was that this animal strongly resembled a reconstruction of an animal postulated by William Beebe (the renowned explorer and scientist who sank into fame by riding the first bathyscaphe down into the Bermuda trench in 1934) as an intermediate between flighted animals like *Archaeopteryx* and more primitive ground dwellers.

Potentially, this has important implications for the beginning of powered flight, first, that flight did not originate on the phylogenetic branch between typical dinosaurs like dromaeosaurs and early birds like *Archaeopteryx*. Instead it was present, perhaps in a primitive gliding form in the common ancestor of both, and then lost (as in the lineage leading to ostriches) in larger later dromaeosaurs. The authors presented three lines of evidence suggesting that *Microraptor gui* was a flighted animal. The presence of a large airfoil type structure on the hind limb (as in the specimen that I described with Gao Ke-Qin and Ji Qiang), the orientation of the thigh bone, which may stick straight out of the hip to support the airfoil (as in a flying squirrel), and the presence of asymmetric feathers. Asymmetric feathers were first

Kite flying is a major activity in China and people of all ages and walk's of life partake. Here a lifelike raptor kite patrols the air above Tiananmen Square.

Detail of the feathers on the back of Dave's arm. Notice the distinctive herringbone pattern suggestive of the vane of a feather.

reported in *Archaeopteryx*, and it is their distinctive shape that confers an aerodynamic function in extant birds that has been used as evidence for powered flight.

All three of these things correlate with flight, but they do not necessarily indicate it. On further inspection, the lateral orientation of the hind limb is suspect. This underscores one of the difficult problems and frustrations in paleontology—that of studying behavior. The animals that have to do with the origin of flight have been dead for over 100 million years. We cannot observe them directly; instead, we need to look for artifacts, basically byproducts, of a certain behavior that can be preserved in the fossil record. Such byproducts, like trackways, or what we call fossil "snapshots" that preserve behavior displays (such as the nesting dinosaurs I described years ago from Mongolia), are few and far between. None has been discovered, and can hardly be imagined, that would display flight.

In a News and Views piece accompanying the *Microraptor gui* paper, Kevin Padian pointed out that, while all these things are suggestive of flight, they are at best only correlates with it, and still can not be shown conclusively to indicate flight. But even if *Microraptor gui* did not fly, the "dinosaur biplanes," as they have been dubbed, may say something very interesting about the origin of the avian wing. Did wings first appear as a serial structure on both the fore and hind limb? Beebe had postulated such a configuration, based on genetic anomalies in modern birds that occasionally produce wing-like structures on the hind limb and on his ideas about the origin of flight originating through a gliding stage. Perhaps for gliding, or perhaps for some other function, such as camouflage against the trunk of a tree, an interesting idea to be pursued is that wing-like structures were originally present on both fore and hind limbs, and then, with the onset of powered flight, the forelimbs took on the primary role of aerial locomotion.

Two hypotheses have been proposed about the origin of flight, the ground up and the trees down.

The ground-up hypothesis came as a corollary to the theory of a theropod ancestry of birds, advocated in its modern form by John Ostrom in the 1970s. This theory says flight evolved in small theropod dinosaurs in a scenario where fast-running dinosaurs were able to propel themselves forward, extending their leaps by flapping their feathered arms. The trees-down hypothesis is pretty obvious—that flight arose in small theropods through an intermediate gliding stage.

Lots of evidence has been marshaled for and against each of these ideas. Unfortunately, few rigorous tests have been developed, and often deep-seated, religious-like belief in one or the other has been used to address the question of how flight originated. In reality, we know little. Both of these theories as constructed are notoriously hard to test. Nevertheless, a better understanding of the mechanics of flight in modern birds is starting to give us a better understanding of how it may have evolved. Work by Ken Dial of the University of Montana has shown that flapping in some modern birds helps them to accelerate while running on steep inclines or even near-vertical surfaces. While this may be taken as evidence for the ground-up hypothesis, other evidence suggests the trees-down one. An expansive look at the data shows consilience between the two theories, which somewhat changes the question from the simplistic trees down or ground up. In the words of Yale ornithologist Rick Prum "The main question now is: Which components of avian flight evolved in the context of terrestrial locomotion and which evolved in the context of aerial flight?"

What we have learned from the feathered dinosaurs of Liaoning is not so much about how flight evolved, but more about how it didn't. Namely that feathers, which Alan Feduccia has described as "magical structures," are decoupled from the origin of flight and were parts of dinosaurian history long before the evolution of birds. Arguments about dinosaur biplanes, whether they flew or glided, and what this means about the origin of flight are, however, just beginning.

Xu Xing and I have recently published a review paper on theropod dinosaurs for *Annual Reviews of Earth and Planetary Sciences.* In this review we list nine kinds of feathered dinosaurs known from Liaoning. Obviously, this list will be added to

Surrounding the tail and at the back of the skull indications of protofeather were found with a specimen of *Dilong*. These extended about 3 cm from the body.

shortly; however, our latest discovery, *Dilong paradoxus*, merits special mention.

Throughout this book, I have hinted at the possibility that *T. rex* had feathers. Although this is something I would have expected, how surprised I was when, in the spring of 2004, Xu Xing showed me a skull of a beautifully preserved advanced theropod. Immediately, the shape of the hole behind the eye socket and the d-shaped front teeth indicated that it might be a tyrannosaur. Further analysis showed this to be correct. But there was another specimen, a specimen that was more fragmentary but clearly the same species of animal. The first specimen, from the lower fluvial beds, was remarkably complete and three-dimensional. The second, from the paper shales, was smashed flat and fragmentary. But along the tail and behind the head there was the unmistakable presence of protofeathers—the same sort as in *Sinosauropteryx*.

What does this mean for tyrannosaurs? Because *T. rex* is descended from the same ancestor as *Dilong*, *Sinosauropteryx*, and all the other feathered dinosaurs and avians, we would expect it to be feathered. Does this mean there were feathered 12-meter animals looking like giant roadrunners? Probably not. If feathers originally evolved for insulation, animals would have come into thermal balance with their

The skull of the "paradoxical emperor dragon" *Dilong paradoxus*, a primitive tyrannosaur from Liaoning. This animal is much smaller than *T. rex*, and has the primitive condition of three fingers on the hand instead of two as in its famous relative.

environments at a certain mass. Large mammals like elephants, for instance, need to dump heat rather than hold on to it and, consequently, are nearly hairless. Back to what this means for *T. rex*.

When a *T. rex* chick hatched, it might have weighed only a kilogram. If warm-blooded, the discovery of *Dilong* and other feathered dinosaurs indicates that the chick would have had at least a coat of protofeathers. As it grew, this covering would have been lost, perhaps only present on parts of the body as an adult.

A radical departure from the status quo? Not really, just something many of us hypothesized long ago that is in the intermediate zone of testing, and just another testament to how discovery of the Jehol fossils is shaking things up worldwide.

CHAPTER 9

The Birds

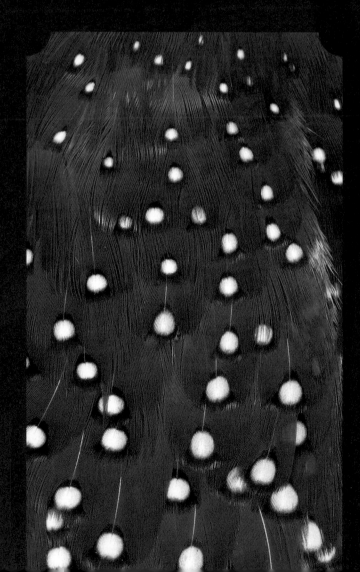

The Immortal Emperor, Karaoke Kings, and Hens' Teeth

There are more species of birds, or, better yet, avians, than of any other terrestrial vertebrate living today. In size, they range from Cuba's tiny bumblebee hummingbird, which can perch on a penny, to Africa's imposing ostrich, which can weigh up to 350 pounds. The Malagasy elephant bird, or *Vouron patra*, which was hunted to extinction at about the time Europeans arrived in the 16th century, was even bigger, at about 1,100 pounds. As variable as avians are in size, they are also structurally diverse—from giant gliding aerialists like condors and albatrosses to penguins with torpedo-shaped bodies. Avians are so diverse that there is not a beak, claw, or body form they have not adopted, and hardly a food source on the planet they do not utilize.

Yet we know very little about how diverse in terms of size, form, or habit avians were in the past. The fossil record of birds is pitiful, so bad that before fossil birds were unearthed in the Liaoning beds, fewer than 20 decent avian specimens had been recovered from the entire 180 million years of the Mesozoic. Now there are hundreds, if not thousands.

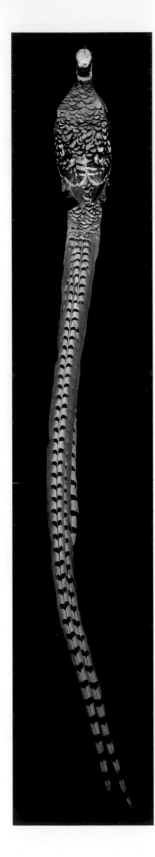

The best-studied Mesozoic avian is *Archaeopteryx lithographica*, discovered in the mid-19th century in the Late Jurassic limestones of Bavaria. The only bird found in the Solnhoffen sediments, it is about 135 million years old and is a perfect morphological intermediate. On one hand, its features are a mosaic, reminiscent of nonavian dinosaurs, with a long tail and teeth; on the other, they are basically identical to those of modern birds, with elongate forelimbs and dramatic feathers. Compared overall with modern birds, *Archaeopteryx* was very primitive, so no one expected that the avians in rocks in Liaoning that are only slightly younger would be so modern in form and diversity.

The skies over Northern China at that time were alive. Apart from the pterosaurs in the Liaoning sediments, more than a dozen birds have been described. Some of them are primitive types, with teeth and long tails, such as *Jeholornis*, while others, such as the pigeon-sized *Confuciusornis*, whose fossils are ubiquitous at Liaoning, are more modern in appearance. None of these avians, however, is closely related to any living species. Nevertheless, the diversity of body form is surprisingly reminiscent of modern bird communities. For instance, in some modern birds, such as the bird of paradise, the males have long, brightly colored tail feathers. In addition to the typical bird feathers, some *Confuciusornis* specimens show a pair of long tail streamers. These are not aerodynamic structures, and they are not present on all the specimens that have been found, which highly suggests that they are display feathers particular to only one sex.

The tail feathers of a pheasant. Such feathers can grow nearly three meters long and are used as integral parts of costumes in the Chinese opera. Another example of how feathers are modified for several reasons besides flight.

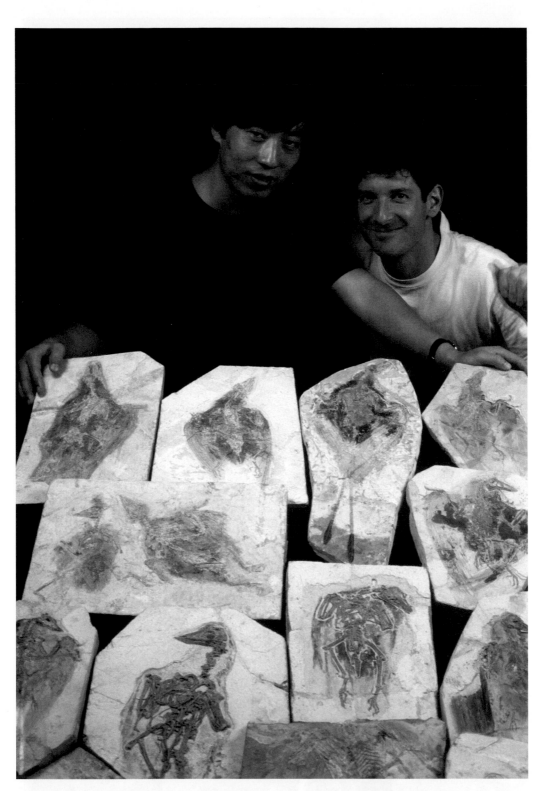

Hundreds of specimens of *Confuciusornis* have been collected. Here Ji Shu-An and Mick pose in front of several tasty examples.

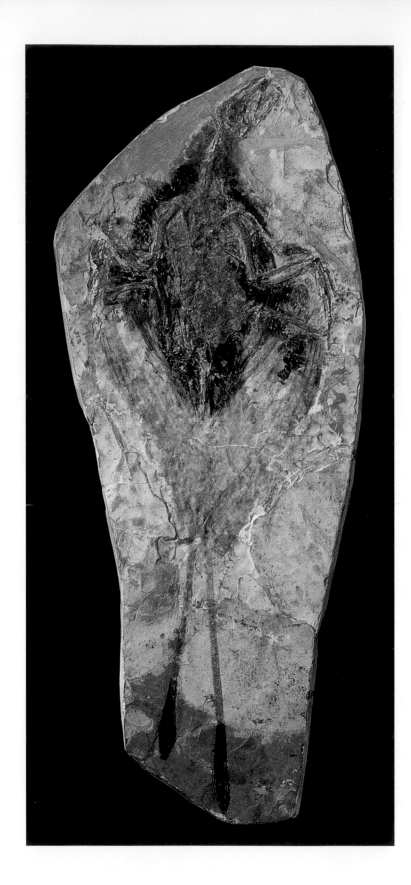

A great specimen of *Confuciusornis*, showing the extremely long tail feathers reminiscent of a Bird of Paradise. The detail shows how these feathers expand toward their tip. Because such specimens are found only rarely, some have suggested that they represent males and perhaps appeared seasonally.

The mosaic of structures found in *Confuciusornis* reminds me of the mythological *Feng Huang*. Often mistakenly referred to as the Chinese phoenix, it is known in the West more for adorning the covers of Chinese take-out menus than as the immortal emperor of birds who rules over the southern quadrant of heaven.

According to legend, *Feng Huang* first appeared to the Yellow emperor, Hung Di, more than 4,000 years ago. Like many mythical creatures in Chinese culture (and some fossils from Liaoning), *Feng Huang* is a composite of several birds. It has the head of a rooster, the body of a pheasant, and the feathers of a peacock. It also has both male and female attributes. *Feng* represents the male, or yang, aspect of the bird and is aligned with summer and the solar cycle. *Huang*, the female side of the bird, represents the lunar cycle and was a symbol of the empress. The *Feng Huang* appears in times of prosperity. According to legend, its singing voice formed the basis for the five-toned Chinese musical scale.

This description of the emperor of birds leads to my friend Luis Chiappe, for no description of the birds of Liaoning would be complete without some mention of him. Luis Chiappe is an expert on avian paleontology at the Natural History Museum of Los Angeles County. We have worked together a long time and frozen

Luis Chiappe, friend, eminent avian paleontologist and a cultured man.

our asses off together in the Mongolian deserts, done copious research, and happily sung ourselves hoarse in odd bars on the other side of the world.

During the first few years of work on the Liaoning fossils, Luis, Mick and I spent significant time together in the Middle Kingdom. Good traveling companions are as necessary to thriving far from home as is your own roll of toilet paper.

Apart from his extensive abilities as a "serious" scientist, Luis is passionate about life and has a retinue of life experience. Although just over 40, Luis hasn't missed much. As traveling companions, Mick and Luis are a study in contrasts. Mick likes beer, Luis wine. Mick is egoless, Luis, let us say, is an Argentine, a Porteno (a Buenos Aires native, notorious for bravado and arrogance) at that. Luis loves music; Mick plays it. Luis is an aristocrat; Mick is an Irish-American from Appalachian Upstate New York. Occasionally, their humors and passions converge, creating memorable chemistry.

Karaoke is a common diversion in China and throughout Asia. Some even attend karaoke schools so they can have a couple of great performances all ready to go.

Judging from the number of karaoke establishments (which run from totally legitimate to obvious fronts for more intimate activities), karaoke is the most popular form of musical entertainment in China and many other parts of Asia. One evening in Beijing after a long day at the National Geological Museum, we decided to visit the karaoke lounge upstairs in our hotel. We had just wrapped up photographing the last images for what would become a lengthy monograph on *Confuciusornis*, still, I'm proud to note, considered the definitive treatment. We were feeling good about the work we had accomplished that spring day.

After a round of drinks, we were presented with the song catalogue. In some places the catalogue is the size of a telephone book, with extensive coverage of Western and Chinese pop, rock, folk and traditional music. Western music is popular in China, and often we are asked to sing the locals' favorite American songs. This night was no exception. Two young women sitting at a table next to us, Pe Chong and Tong Ning, made our choice for us as they excitedly pointed to a tune called "Casablanca." We had never heard of it, but what the hell, the TelePrompTer is there, so it can't be that hard, right?

Our table, which was in the shadows at the back of the room, was called, and Mick and Luis took the stage. The music started, and although they could read the words, like Karaoke singers round the world, they just couldn't follow the tune. The packed house was laughing, at first politely, then hard. It was quite a different scene from when Mick sings alone and people come over to our table to compliment his talents. Beijing people love Elvis, and any tune by the King always results in a round of drinks for our table. Luis and Mick made their way back to where I was sitting. Embarrassed, I tried to disassociate myself from the perpetrators of the worst karaoke performance I had heard up to that time, by talking to a Taiwanese business-man at the adjacent table.

Luis, in an effort at redemption, decided he would sing "Bessame Mucho." Mick had already been humiliated once, so he was not going to go up there with Luis again and told him so. A confident Luis proclaimed, "Of course I will do this song by myself; I am always on my own!" Besides, as we were assured, this song is so popular in Latin America that everyone knows it. Luis took the stage and the lights came up. He inhaled mightily for the first verse, and got ready to let it go. The TelePrompTer flickered, and the words appeared against the usual campy background of blond white people at the seashore (unrelated Eastern European B roll is a common thread throughout the Karaoke world). Not only was the tune a little different from the one Luis remembered from his youth, the words on the screen were in Chinese characters, rather than English or Spanish. Luis stood there with the music blaring and his mouth open.

Everyone else in the room was dead quiet. Luis just began to bellow, "Mick, Mick, come here." I don't know what he was thinking, but Mick popped up to join Luis on stage. The two of them just stood there, saying nothing until the song ended. No one clapped. No one said a word. Pe Chong and Tong Ning were embarrassed to have met us; we had defiled a great tradition. The three of us headed for the door.

Luis may be a terrible singer, but he is an excellent scientist. He has done an outstanding job of placing myriad primitive birds into a phylogenetic context. We have worked together on several projects, including our *Confuciusornis* monograph, and Luis has worked with Chinese colleagues on a number of other Liaoning birds.

First described in 1996, *Confuciusornis* provides the intermediate link between *Archaeopteryx* and more modern avians. *Confuciusornis* is a hodge-podge of primitive and advanced characteristics. Like *Archaeopteryx* and nonavian dinosaurs, *Confuciusornis* has three fingers that are not fused into a single element, and the bones of the feet are all separate. The tail, however, is shortened to form a short pygostyle, which is an element composed of several tail segments fused together. *Confuciusornis* is toothless and has indications of a beak. The shoulder girdle resembles that of modern birds, suggesting that it was a much more capable flyer than *Archaeopteryx*. While *Archaeopteryx* looked much more like a nonavian dinosaur, here in the skies of China over 120 million years ago, we have a highly developed modern-looking avian.

Living birds all share several unique characteristics, features that are found in birds and in no other member of the animal

Confucius and Confucian principles have a big imprint on the traditional culture. One of the precepts is respect for authority.

kingdom. Their presence is evidence that birds are descended from a common ancestor. A lot of the characteristics by which we define modern birds are in *Archaeopteryx*, which, while it is looked at as the link between typical dinosaurs and typical birds, is in many ways very un-bird-like. It does have the bird-like presence of a wishbone and a first toe that points backward. No nonavian dinosaurs have a reversed first toe, but many (including *Tyrannosaurus rex*) have wishbones. The main reason everyone has pointed to *Archaeopteryx* as a close bird relative is that it has feathers. As pointed out earlier, for a long time, it was believed to be the most primitive animal to have feathers, and because of them, it was considered an avian from the outset. *Archaeopteryx* is, however, so primitive that one specimen was mislabeled as a

The three fingers of the hand of Dave correspond to the thumb, index and middle finger. They are still obvious in modern birds although they are fused with many wrist bones into a compound element.

pterosaur until long after it was collected, because, unlike most of the better-preserved *Archaeopteryx* specimens, it did not have distinctive preserved feathers with its fossilized bones.

A way to understand the differences between *Archaeopteryx* and modern birds is to compare it with your next chicken dinner. Of course, it would have to be a Chinese chicken dinner because those come with the head and feet still attached, but you can just use your imagination. Starting with the head, instead of a toothless beak and bill, *Archaeopteryx* had a mouth full of small pointed teeth. Associated with modern powered flight, the forelimbs and shoulders of living birds are highly modified. Most of these modifications include a stiffening and reorientation of the shoulder bones so that the arms extend out from the body rather than lateral to it. On the extremity of the arm, the wrist bones are combined, and the finger bones are fused into a single structure. In *Archaeopteryx*, the shoulder and forelimb are just like those of a more primitive dinosaur like *Velociraptor* or *Sinosauropteryx*, where the hands are composed of three separate fingers.

In the foot and the tail, the story is the same. In *Archaeopteryx*, the tail has around 20 segments and is nearly as long as the animal itself; in living birds, the tail is reduced. What my grandmother called the part that cleared the fence last is a small triangular piece of flesh formed of a few tail bones, which supports the tail feathers. This you can see on just about any rotisserie-roasted chicken in the West.

As anyone who frequents dim sum parlors knows, the foot of a chicken is composed of one big bone and lots of little ones that form the toes. The big element, called

Even very tiny birds, like this specimen of "*Liaoxiornis*" which is less than 10 cm long, are preserved in the Jehol sediments.

the tarsometatarsus, is a composite formed by the fusion of the three separate foot elements in *Archaeopteryx* and typical advanced theropod dinosaurs. Especially when examined in the context that feathers were present in lots of animals other than birds, these differences between modern avians and *Archaeopteryx* give a strong impression that if you could bring all of them to life, the first "bird," *Archaeopteryx*, was a lot more similar to nonavian dinosaurs than it was to truly modern avians like pigeons and ducks.

The stomach area of *Jeholornis* preserving its last meal of seeds. This specimen was found in the Jiufotang Formation and provides the first evidence of seed eating in avians.

The importance of the Liaoning birds isn't just the epiphany brought on by the discovery of *Confuciusornis* that the anatomy of modern birds was part of an evolutionary assembly process, as opposed to being rolled out wholesale. The diversity of specimens is most surprising. Over 20 new species have been discovered, and they are noteworthy both for their relationships and their anatomical differences. And there will certainly be more to come.

Both large and small birds have been found in the Liaoning beds. Purportedly, the smallest bird is *Liaoxiornis*, at only 82 millimeters long. Specimens of it are very common; however, it is now considered to be a juvenile of another very common mature bird. Other fossils have even shown that the front half of the "*Archaeoraptor*" chimera described earlier is in fact a true avialan, recently named *Yanornis*. Stomach contents indicated that it was a fish predator, and its long-toothed snout was certainly fitted for this use. Another specimen indicates that some Jehol species were seedeaters, as fossil seeds have been found in the body cavity of *Jeholornis*, a recently described bird from the younger Jiufotang Formation. Unfortunately, these seeds cannot be referred to any known plant group by paleobotanists, but because they were ingested whole, it seems that *Jeholornis* processed them in the gizzard, as do modern birds.

Sapeornis and *Jeholoronis* are the largest Liaoning birds yet discovered. These were large animals, with wingspans of just over two feet. Yet, these were not the largest animals to patrol the ancient Liaoning skies, by far. Even the largest birds were dwarfed by some of the Liaoning pterosaurs I talked about earlier; they had wingspans five times this size.

The most advanced birds found in the Liaoning deposits are the enatiornithines. "Enantiornithes" means "opposite birds" and they are named that because aspects of the anatomy of their feet are formed in the opposite way from other birds.

In Luis' mind, he and Mick are also opposites, a distinction that is catalyzed when they are in China. Luis loves art; Mick's interests are self-evident. Whenever we can pry ourselves away from work, we check out the galleries, museums, and shops. Many hotels in Beijing exhibit art. The Shangri La Hotel is located in the city's northeast section. It is an elegant hotel, but decorated with the usual hotel art

Some of the great stories of Chinese mythology are the stories of the Eight Immortals. Each immortal is imbued with patronage of life, literature, medicine, actors, flowers, home, music or happiness. Each also has particular skills and attributes, like a jade flute or jade tablet, a paper donkey or a palm leaf. Here they get together for a few drinks. When attacked during such a gathering, legend has it that they created an impromptu *Kung Fu* style on the spot—now known as Drunken Fu.

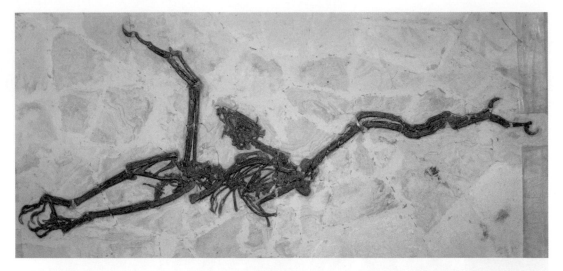

Sapeornis, the largest of the Jehol birds yet discovered. About the size of a small sea gull, the long wings would have made *Sapeornis* a powerful flyer.

all too familiar to any international business traveler. We retired to the Peacock Lounge for an aperitif. I left to take a leak, and Mick went out to admire the hostesses in the piano bar in the main lobby.

We converged in the lounge to find Luis standing in the middle of the room gesticulating like a madman. He was staring at a big picture, throwing his arm in the air in great curved motions, tears running down his face, punctuating each movement with a dramatic flick of the wrist and mumbling, "He is such a master," over and over. Mick and I thought the picture was butt-ugly and started laughing hysterically. Luis was pissed. Being viewed as something of a Philistine, I was spared. Mick was not; we had both laughed at Luis too hard and for too long. He had had enough. He went into a speech that is some weird hybrid between Richard III and a character out of the Marx Brothers' movie, Animal Crackers.

Looking right into Mick's eyes, Luis proclaimed, "I am not like you; I am a sensitive, cultured man. Look at my eyes, look! Look at these tears! I am not afraid to cry, but I can also be very hard!" If Luis thought that we were going to take him seriously, he was wrong. I laughed so hard at the Argentine *Feng Huang* showing both yin and yang sides that beer came out my nose. Mick was hyperventilating. Just as with Karaoke, Luis grew silent and defeated.

The focus of most of Luis' contemporary research, the opposite birds were the most successful radiation of avialans in the Mesozoic era, and they all disappeared at the end of the Cretaceous, coincident with other nonavian dinosaurs. Enatiornithine birds are very common in the Liaoning beds, and even the first avian specimen found there, *Sinornis*, belongs to this group.

When the Liaoning birds are collectively placed in a systematic context, strange things happen. Some of the long-tailed and toothed varieties like *Sapeornis* and *Jeholornis* are more primitive than the beaked and short-tailed *Confuciusornis*. Nevertheless, skeletal modifications to the arm, or wing, show that these animals were powered flyers, much more like modern birds than the *Archaeopteryx*, with its extremely primitive and dromaeosaur-like flight apparatus. More closely related to modern birds are the enatiornithines, followed by several species of unusual birds, some of which are ground dwellers, others penguin-like, vestigial-winged, toothed divers, and others small toothless flyers. None of this later group is from Jehol; consequently, all of the Jehol birds are somewhat removed from the great radiation of today's birds.

Yet even with this small sample of animals at the base of the family tree, unexpected things are happening. As with many evolutionary things, features come and go. Avialan genealogy indicates that primitive birds have teeth. This can easily be demonstrated with animals like *Archaeopteryx*, and *Sapeornis*. These teeth are vestiges of their close relationships to toothed dinosaurs like dromaeosaurs and troodontids.

Confuciusornis is more closely related to modern birds and has a toothless beak. Enatiornithines are even more closely related to modern avians, and many lack teeth, but then things get weird. Teeth reappear independently in myriad forms, including enatiornithines and even some avians that are very closely related to modern species. Luis Chiappe indicates that teeth have appeared independently several times in avian evolution. How can such parallelisms be explained? Like feathers, "evo-devo" experiments are starting to bear on this problem. In a 2003 paper in the *Proceedings of the National Academy of Sciences*, a group of European scientists said they were able to grow teeth on the jaws of chicken embryos by transplanting cells of mice to developing chicks. What these experiments showed is that the genetic machinery to grow teeth is still present if, according to the authors, "there are cells that are capable of responding to it." In this light, the occurrence and recurrence of teeth can be seen more as simple switching of genetic developmental switches than a simple linear process of progressive improvement.

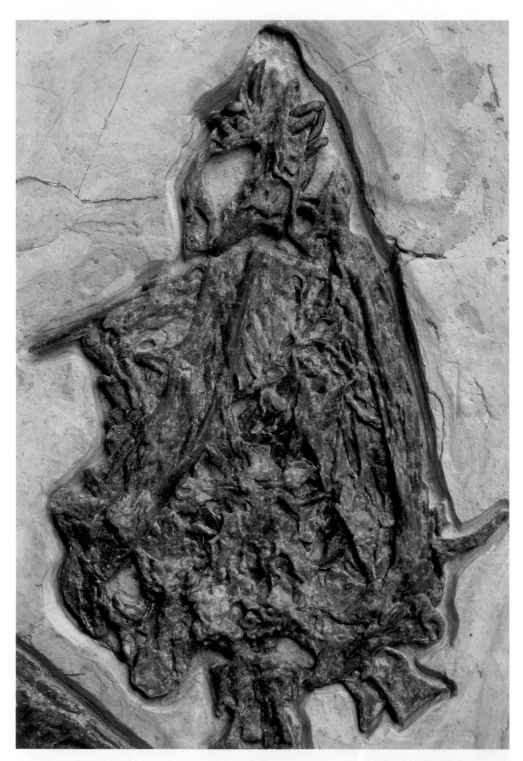

A detail of the skull of *Sapeornis,* showing that all though aerodynamically advanced this avian still retained teeth.

Still, no birds have been found in the Jehol deposits that are part of the great radiation of modern birds. There is no bird that is preferentially more closely related to any modern bird than modern birds are to each other. Although controversial, some recent evidence, analyzing the DNA of living birds and using the data as a molecular clock, has projected the diversification of modern avians to times that are roughly commensurate with those represented in the Liaoning rocks. Other anecdotal, but again controversial, biogeographic evidence also suggests that bird lineages are old.

So where are they? That's a good question, and we are left with two alternatives, that they were there in the Early Cretaceous and we have just not found them, or that they were yet to evolve.

The diversity of avians at Liaoning continues to multiply. By the time you read this, the list will already have grown. This is great, for a lot of reasons. One is that it shows just how perversely bad the fossil record is. Up until 10 years ago, you could fit every single specimen of Mesozoic avialan ever collected on my desk. Granted, I have a big desk, but when you consider the thousands of species that must have been present, and at least 135 million years that need to be accounted for, this is pitiful. Second, the sorts of fossils that have been found only bolster support for the theory that birds are descended from other dinosaurs, a theory developed in the absence of this evidence.

A Beijing duck dinner would be a fitting celebration. Let's not ask Luis to sing.

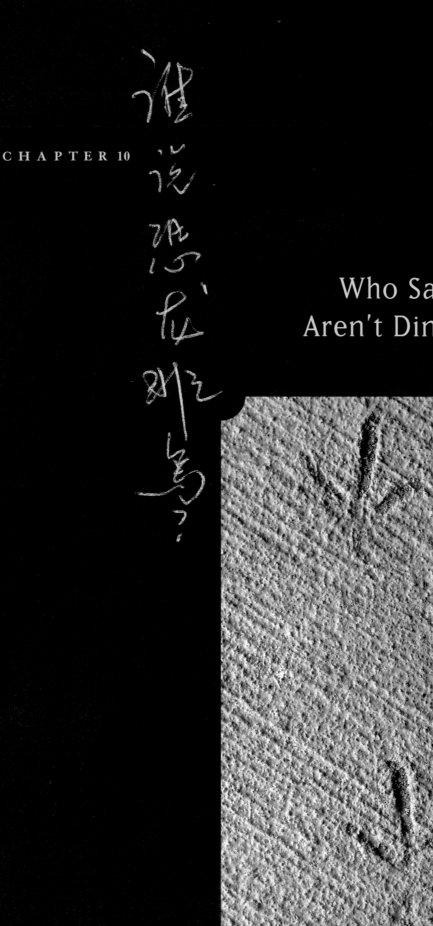

CHAPTER 10

Who Said Birds Aren't Dinosaurs?

"The Greatest Embarrassment of Paleontology in the 20th Century"

In November 1996, Alan Feduccia was quoted in *Science* as saying, "The theropod origin of birds will be the greatest embarrassment of paleontology in the 20th century." It was not an even-tempered comment. Science is not always a genteel world of polite discourse. With so many specimens of great fossils that directly bear on the high-profile controversy over bird origins, perhaps it should be no surprise that things got nasty. At the center of the controversy is the single question of whether modern birds find their evolutionary origins in dinosaurs. While the proposition that avians are evolved from dinosaurs is the focus of heated debate today, it is not a new idea.

The first to propose the theory that dinosaurs and birds were related was Thomas Huxley, a major figure in Victorian paleontology. A skilled anatomist, he was a fierce proponent of Darwin's theory of evolution via natural selection. He was also one of the first to do detailed anatomical studies of the skeletons of modern birds, which he eventually published in two papers in 1868 and 1870. Shortly after completing his studies of

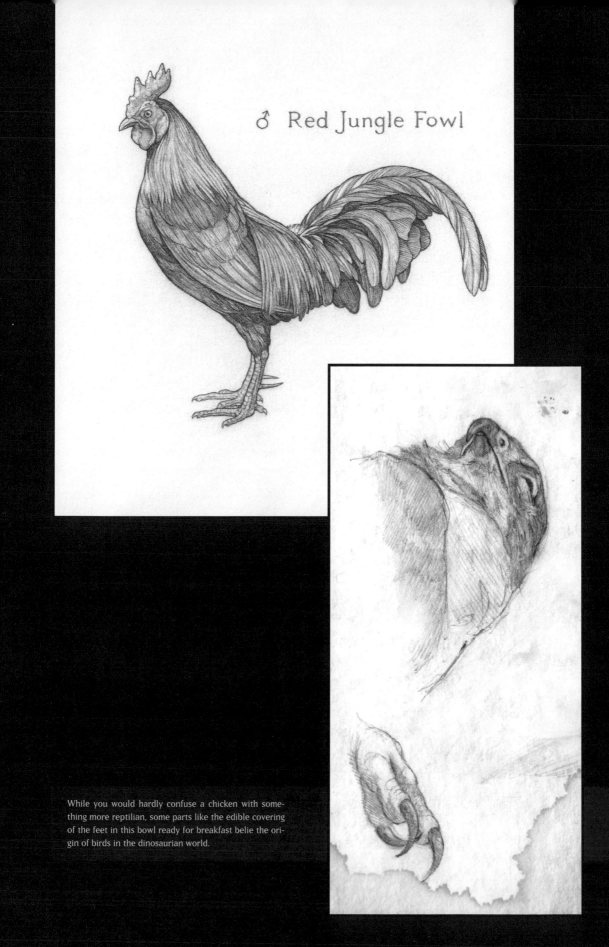

♂ Red Jungle Fowl

While you would hardly confuse a chicken with something more reptilian, some parts like the edible covering of the feet in this bowl ready for breakfast belie the origin of birds in the dinosaurian world.

birds, Huxley was shown fossils of the carnivorous dinosaur *Megalosaurus*, from the English countryside. He recognized that the bones of this nonavian theropod dinosaur were nearly identical to the *Archaeopteryx* specimen that the British Museum had acquired in 1862, and also to modern birds.

Before the Royal Society in 1863, Huxley delivered his most famous paper, proposing that birds find their origin in dinosaurian reptiles. As evidence, Huxley presented a list of 35 features shared by nonavian dinosaurs and birds. Of these 35 features, 17, such as the presence of hollow bones and elongated neck bones, form the beginnings of the contemporary data set that empirically demonstrates that living birds are descended from dinosaurs. Nearly every contemporary dinosaur paleontologist in the world accepts Huxley's theory of dinosaurian origins for birds. But there are a few dissenters.

The discovery of the feathered dinosaurs should have lain to rest any lingering doubts about the link between nonavian dinosaurs and avians. In fact, many in the academic community, especially ornithologists (scientists who study living birds and are largely unfamiliar with the paleontological literature) changed their minds, as all scientists must occasionally do, and recanted their previous support for alternative theories. Nevertheless, right after the first Jehol specimens were collected, a few paleontologists, who with Taliban-like zeal believe it is not possible for birds to find their ancestry in theropod dinosaurs, were quick to reject this new evidence.

Some of this group formed a loose association that has been given the acronym BAND (Birds Are Not Dinosaurs). BAND adherents have developed a cottage industry of nay-saying and incredulity. Not large in number, BAND members are noteworthy for their obstreperous "evolving" rhetoric and their failure to observe modern scientific methods.

Some of my most depressing days are the ones when I have to deal with the feeding frenzy of misinformation and bad science chummed up by BAND around almost every discovery that is made concerning bird origins and feathered dinosaurs. It is not as if these guys are stupid, or that on occasion BAND members have not made contributions. Yet, most of their comments have nothing to do with science. I feel as if many of the papers written by BAND members should carry the prelude to one of the television shows I grew up with: "There is a fifth dimension beyond that which is known to man. It is a dimension as vast as space and as timeless as infinity. It is the middle ground between light and shadow, between science and superstition, and it lies between the pit of man's fears and the summit of his knowledge. This is the dimension of imagination. ..." *The Twilight Zone,* however, displayed more variety in its creativity.

The most outspoken members of BAND include Larry Martin, a paleontologist from the University of Kansas and one of the four Americans on the Academy of Natural Science's panel that announced the "discovery" of the Liaoning fossils; Storrs Olson, a Smithsonian Institution ornithologist, whose research focuses on fossils from the recent past and who sat with me on the panel investigating the *Archaeoraptor* affair; Alan Feduccia, an ornithologist from North Carolina State University; and John

Here a an artist touches up her canvas. While not making a value judgment, science differs from art in that there it is more about evidence than about undirected interpretation. Science is at its best when constrained by rules, a case when art is at its worst. These differences are unrealized by some.

Ruben, a zoologist at Oregon State University.

Their response to the discovery of fuzzy and feathered dinosaurs in the Liaoning deposits was not untypical of their methods. Initially, the claim was made by some of the BAND members that the "dinofuzz" of *Sinosauropteryx* is not a feather-like external covering, but internal structures that supported a crest, as seen in iguanas. In the words of Feduccia, "What you see is a darkened area from the nape of the neck to the tip of the tail. ... And it's almost certainly one of those lizard-like frills running down the back. It has nothing to do with feathers." Feduccia had not, of course,

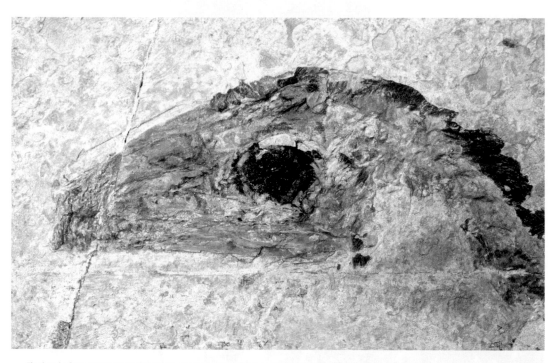

The head of *Sinosauropteryx*, showing the distinctive covering of protofeathers on the nape of the neck.

The integumentary covering on the back of *Sinosauropteryx* has been interpreted by some to be the same as the frill on the back of this iguana. On close inspection the covering of *Sinosauropteryx* can be shown to have covered the entire body, not just a Mowhak down the back.

examined the actual specimen before commenting. Neither had Ruben, who claimed he could determine from photographs that the lung structure of *Sinosauropteryx* was extremely primitive, obviating any chance that this animal was warm-blooded.

Living birds have highly modified lungs, where air flows though the lung in one direction. Often the lungs themselves actually invade the hollow cavities of the bones. Ventilation of the lung occurs by a rocking motion of the thoracic cavity. Unlike mammals, there is no separation of the trunk cavity by a diaphragm. Ruben's team concluded that there was a separation present in the thoracic cavity of *Sinosauropteryx*; therefore, it did not have an avian type of lung. Note that having an avian lung is immaterial, because avian respiration (unless specially created) had to have evolved from a more primitive system of two chambers separated by a diaphragm.

Ruben's evidence for a two-chambered body was a dark stain on the fossil, which he interpreted to be the residue of the liver. This stain seemed to have a very sharp front border, suggesting to Ruben that the body cavity was separated by a diaphragm, as in mammals and crocodiles.

It was funny when Canadian paleontologist Phil Currie (a co-author with Ji and me of feathered dinosaur papers) pointed out that the *Sinosauropteryx* "liver" was just a smudge of carbon and glue on a rock and that the front margin was an artifact of the way flakes of rock entombing the skeleton had split.

One would like to think that the description of *Caudipteryx*, a small dinosaur with feathers exactly like those of modern birds, which I published with colleagues in *Nature* in 1998, would have silenced these critics. Because the feathers were irrefutable, the critics changed direction and claimed that *Caudipteryx* was an avian, not a dinosaur; after all, it had feathers. Right?

BAND spun this one better than a presidential political action committee. In a contentious article published in *Nature* in 2000, Terry Jones (an ex Ruben student), John Ruben, and colleagues tried to make the point that ratios determined from the trunk and limb lengths of *Caudipteryx* more resemble those of modern ground-living birds than extinct dinosaurs. In the paper itself, they did not indicate that this was evidence for *Caudipteryx* being a flightless bird. They saved that for the press release—a favorite technique. On closer scrutiny, it was difficult to replicate their analysis, as several of the measurements they reported were either inaccurate, from bones that were so heavily restored as to render any measurement imprecise, or based on bones that do not exist. They may have stated that they relied on others' reconstructions, but there is no excuse for not verifying these with the original description, especially when the

Caudipteryx zoui, an unusual flightless feathered nonavian dinosaur. *Caudipteryx* is one of the more common feathered dinosaurs and this specimen has a stomach full of gastroliths perhaps indicative of an herbivorous diet. The inset shows details of the feathers on the hand, entirely modern in construction, except for their symmetric vanes.

reconstructions appeared in a popular publication. On the strength of their press release and the media's taste for controversial stories, several media outlets picked this up as definitive evidence against the presence of feathered dinosaurs.

In response to what we viewed as a disingenuous piece of work, Dublin paleontologist Gareth Dyke and I crafted a reply to *Paleobiology*. The reply received strong reviews, yet, as the journal's protocol dictates, authors are given the opportunity to reply to criticisms of their work. Ruben et al. never replied, and the paper languished on the editor's desk and only recently appeared in *Acta Palaeontologica Polonica* in early 2005.

The nail in the coffin of BAND should have been our discovery of Dave and the discovery of unquestionable feathers on the related dromaeosaur, Chong, in 2001. Instead of accepting this, BAND members tried to cast doubt on the authenticity of the specimens. Martin announced, "Researchers have been duped before by elaborate fakes such as *Archaeoraptor*, and it is important that the fossil is a true dinosaur and not an elaborate fake."

One of BAND's major criticisms has little to do with the data and is about its members' refusal to use modern systematic methods to discover the family tree of avians and dinosaurs. A quick rundown of these methods is that we use observations (characteristics of organisms, such as the presence or absence of teeth, feathers, or any other attribute) to create a matrix. This matrix catalogues all of the character observations for all of the organisms under study. A computer program then takes this matrix and finds a family tree that requires the least number of evolutionary transformations (such as the origin of feathers). When more evidence is garnered, either through the analysis of additional characters, through the discovery of new specimens, or by pointing out errors and problems with the original data sets, new trees can be calculated. If these new trees better explain the data (taking fewer evolutionary transformations), they supplant the previous trees. You might not always like what comes out, but you have to accept it.

Any real systematist (or scientist in general), has to be ready to heave all that he or she has believed in, consider it crap, and move on, in the face of new evidence. That is how we differ from clerics.

With subtle changes, the current tree of theropod dinosaurs closely resembles that proposed by Jacques Gauthier of Yale University in 1986. Gauthier's tree, the first empirical tree calculated for these animals, indicated that several advanced dinosaurs

A deformed stack of rocks in the Sihetun quarry. Some paleontologists believe that understanding relationships is a simple matter of collecting fossils and connecting the dots through the rock layers with older fossils found further down the hill ancestral to the younger ones. I am glad it is not that easy as this would take the fun out of it.

were closely related—a group which he called maniraptorans. The maniraptors included animals like oviraptors, dromaeosaurs, and troodontids, and birds (including *Archaeopteryx*).

BAND adherents never supported this, and have instead argued one of two points: that dinosaurs are closely related to a group composed of living crocodilians and their extinct relatives, or that they are related to a nebulous phantasmagorical group called thecodonts. BAND's research is not an attempt to show what dinosaurs are related to by proposing and testing alternatives to the theropod origin hypothesis; rather, it is simply an attempt to show that the theropod origin hypothesis incorrect. For instance, Feduccia has said, "There are times when there is insufficient evidence to make the formulation of a hypothesis feasible."

As cogently pointed out by Richard Prum, such statements cannot even be construed to be scientific, because they cannot be tested, and they can accommodate any possible conclusion.

Aside from the obvious flaw of not using modern empirical techniques to make family trees, BAND often points to what its members perceive as three fatal shortcomings in the hypothesis that birds are dinosaurs: the temporal paradox, which states that the relevant fossils are found out of phase and in the wrong temporal order; the apparent incomparability of the typical theropod dinosaur hand and that of a bird, based on patterns of embryological development; and the supposed impossibility of flight arising from the ground up rather than the trees down.

Feduccia claims that the closest dinosaurian relatives of birds (dromaeosaurs and troodontids) occur much later in time than the first "bird," *Archaeopteryx*. He makes the claim that this means the descendants (birds) appear earlier in the fossil record than do their ancestors (advanced theropod dinosaurs). Ergo, it is therefore impossible that birds are descended from such dinosaurs.

If the fossil record were perfect then yes, we would expect more primitive animals to appear earlier in time than more advanced ones, so animals like dromaeosaurs should appear earlier than birds. But this is not the case—hence, Feduccia's point. However, the fossil record isn't perfect. In fact, it is far from it, for a variety of sampling reasons that have been talked about throughout this book. In fact, it is so incomplete that, to me, it would be surprising for us to find everything in order, where we could essentially connect the dots between fossils and have a family tree.

Furthermore, in claiming this, Feduccia demonstrates his ignorance of evolutionary patterns, because he confuses the linear pattern of direct ancestry and descent

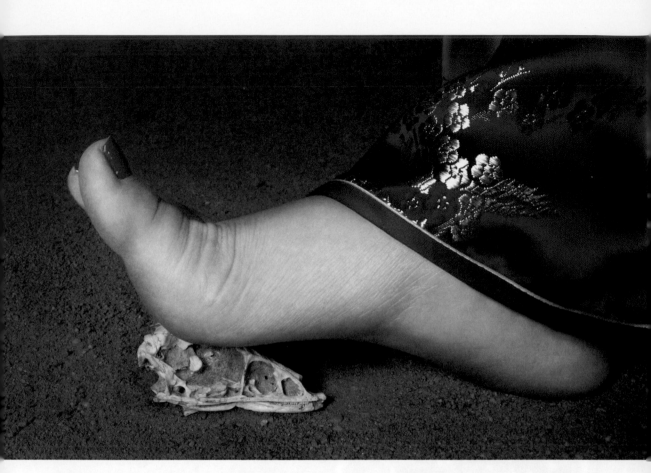

Not all dinosaurs were big. Here is the skull of a small troodontid, an animal closely related to bird origins with a lovely foot for scale.

with the hierarchical branching pattern of genealogy. Although ancestor descendant relationships must exist we are at a loss in recognizing them in the fossil record. Consequently, none of us ever said that any advanced theropod is the direct ancestor of birds, only that dromaeosaurs and troodontids share a common ancestor with birds. If evolution worked in a linear fashion, there would only be a single kind of life today—it would be a linear process of ancestry and decent from amoebas to us. Instead, evolution is a hierarchical process where species divide to form new species, diversity increases, and organisms are related to one another by sharing common ancestors, as opposed to having direct ancestry.

We could easily dismiss Feduccia's argument of a temporal paradox, on sampling grounds alone. Fossils of these animals are very rare, and we would not necessarily expect sampling that was good enough for us to find fossils ordered in the stratigraphic column in a pattern that closely matched the pattern of branching of a family tree.

But trying to develop more empirical tests is always fun. A couple of years ago, Chris Brochu, of Iowa State University, and I expanded on some theoretical work that I did with my American Museum colleagues, Mark Siddall and Diego Pol. Siddall, Pol, and I developed a numerical method that allowed us to test how well a particular family tree matched with the order of appearance of fossils. We analyzed several hypotheses of bird origins, including those preferred by BAND (basically that birds are closely related to crocodiles or to very primitive, poorly known fossil reptile species). What we found was that the theropod dinosaur origin hypothesis fit the fossil record substantially better than any other. Consequently, the only thing paradoxical about this argument is why BAND members still think they have a point about the distribution of bird and dinosaur fossils and the relevance of these data to bird origins.

Three chapters ago we discussed the origin of flight and feathers. We do not need to recap here, except point out that BAND adherents have never quite gotten it that the origin of flight and feathers is decoupled from the origin of birds. To them, if it has feathers, it is avian. They have odd ideas about how all this relates to flight. As Feduccia claims, "It is biophysically impossible to evolve flight from such large bipeds with foreshortened limbs and heavy, balancing tails." Before more discussion of feathers, the point about large bipeds incapable of climbing trees is clearly falsified by the presence of so many small nonavian theropods in the Jehol fauna.

BAND's model of feather origin is entrenched in an old model of feather evolution—the direct transformation of scale to feather—again confusing a linear model of evolution with a hierarchical one. Such a linear model requires that ancestral forms be identified. Usually, these are extremely fragmentary ones that are difficult to incorporate into any hypothesis of a relationship. This brings us to *Longisquama*.

Longisquama is an animal that is known from a few specimens. At least five poorly preserved specimens have been excavated in the Fergana valley of what is now Kyrgyzstan. It is the Fergana valley that produced the famous Fergana horses, so sought-after by Tang dynasty military commanders as war horses. Their large size and heavy features have been memorialized in the equine sculpture of the period.

A tomb figure of a Fergana horse. These horses originated in Central Asia and were the subject of extensive military and artistic interest in Tang Dynasty times.

Longisquama had been known since it was described by the Russian paleontologist Alexander Sharov in 1970. It was not taken seriously by anyone as a candidate for relevancy to ideas of feather origin. A lizard-like animal, it lived in the Early Triassic period, about 245 million years ago. How *Longisquama* is related to other reptiles is unclear. Most who have looked at the problem, however, such as Hans-Dieter Sues of the Smithsonian, find no evidence that it is a dinosaur. It was a weird creature. The unusual thing about it is the presence of long brush-like structures that emanate from its back, hence the name, which means "long scales."

With the evidence of feathered dinosaurs mounting, Feduccia, Ruben, and company grabbed on to this like a lifeboat, suggesting that the long scales had been previously misinterpreted and were actually primitive feathers. They published this in *Science* in 2000, thirty years after Sharov's original publication. By this time, the theropod-bird relationship had gained broad acceptance, and the authors of the *Science* paper describing these findings probably thought that stating that *Longisquama* was a bird ancestor would have been reason for the paper's rejection, as there is not a shred of evidence to support that. In the paper, they just talked about how *Longisquama's* elongate scales could have evolved into feathers. As in other cases, however, this did not stop them from suggesting in a press release: "*Longisquama* would have lived in the right time and had the right physical structure to have been an ancestor—and it was clearly not a dinosaur." Storrs Olson went further. "It's a stake in the heart of the dinosaur theory, which I haven't believed in years."

In the 1920s, a young American named John Blofeld traveled to Beijing to teach English. Quickly, he fell under the spell of the city. He documented his feelings in a book entitled *City of Lingering Splendor*. Blofeld immersed himself in Chinese culture, spending a lot of time with scholars, and enjoyed the more sensual pleasures of the willow alleys around the Forbidden City. He became spellbound with a beautiful young woman, who his compatriots were sure was a *hu li jing*, a vampire fox, able to transform herself between human and animal forms. *Hu li jing* are dangerous, because once under their power, a person's sensibility is lost and judgment clouded. They cause men to do foolish things and keep them under their spells.

In science, as in China, much can be seductive. In the 19th century, a series of influential biologists developed the concept of homology, which is much of the basis for understanding the pattern of evolution. Homology is basically a statement that there is a commonality to parts. In simple terms, this means there is a sameness to the arm of a person and the wing of a bird. Today, we explain this sameness by common ancestry—that is, both humans and birds have forelimbs (wings and arms) because we are descended from a common ancestor that also had a pair of appendages attached to the shoulders.

The embryonic development of structures could be studied, even in the 19th century, and scientists were able to trace the development of adult body structures at a very fine degree of resolution. It is commonly held that the pattern of development is somehow intricately related to the pattern of genealogy.

In birds, it was recognized that during development of the embryonic hand, there is evidence for the early appearance of five digits—a common number among higher vertebrate animals. Living birds have just three digits, and further analysis indicated the final number was determined by the disappearance of two of the initial condensations, which are called digit primordia. If one followed the course of finger development, it became obvious that the two outer fingers (the thumb and the little finger, which we will call fingers one and five, respectively) are lost during a bird's embryonic development. This evidence suggests that the three fingers of birds are homologous to digits two, three, and four of the general vertebrate hand.

When this was first proposed, few well-preserved dinosaur fossils were known, so it was just assumed to be correct. During the past century, however, amazing numbers of theropod dinosaur specimens have been collected that form a series of

In Chinese legend, *hu li jing* are women who have intoxicating effects over men, stealing their power and making them do foolish things. Roughly translated, it means vampire fox, and they have the ability to transform between human and animal form. Mick's interpretation based on years of research.

Qing Qing

The fins of sharks when used as garnishes or prepared into soup are a tasty and opulent menu item. These fins show a distinctive pattern of parallel collagen fibers. Some have argued that it is fibers like these that we are looking at on *Sinosauropteryx,* not protofeathers.

forms, where those with five fingers evolve into those with four, and then into those with three. In opposition to the developmental evidence, the fossil record clearly indicates that the fingers of advanced nonavian dinosaurs like Dave and *Microraptor* are fingers one, two, and three (thumb, index, and middle finger) and not two, three, and four. This presents something of a conundrum for those advocating a theropod origin for birds and is decades-old ammunition for BAND adherents.

Is this a real problem, or is it just a case of *Hu Li Jing*—seduction by powerful developmental evidence? In 1999, Yale scientists Gunther Wagner and Jacques Gauthier proposed a novel idea to explain the so-called discrepancies between what embryology seemed to be telling us and the signal from analysis of the fossils. Gauthier and Wagner pointed out that there is no required correspondence between the earliest stages of finger formation—the digit condensations in the developing hand—and the form and identity of the digits in an adult. They went on to propose that a "frame shift" had occurred, where differentiation into the adult shapes takes place independent of the identity of the finger. While hard to imagine such a frame shift, it is now known to occur in the hand of the living kiwi during development. Like the Hu Li Jing, the evidence from development is not always what it appears to be.

Moving from the three primary objections by BAND, 2003 saw a resurgence of the idea that many of the feathers on the Liaoning specimens are not external integumentary structures, but rather, internal structures. The story goes that these structures are collagen fibers supporting internal tissues. When the animal died, they left a halo of strand-like material around the bodies of the decaying animals. This was first presented as evidence against a fluffy body covering in *Sinosauropteryx*, but has also been used as criticism of a number of Liaoning specimens. There is little substance here, because it was only the specimens of dinosaurs that were questioned; the veracity of the feathers on the avian specimens never raised an eyebrow. They were birds, and they were expected to have feathers.

Subsequent analysis of the *Sinosauropteryx* specimens has indicated that in no way could the fibers associated with the skeleton be collagen, because they covered the entire body uniformly, and they were hollow. Yet this argument was rekindled by South African paleontologist Lingham Soliar, who was surprised when he dug up a dolphin that he had buried two years previously, to find what looked like a halo of collagen fibers around the skeleton. It reminded him of the feathered dinosaurs, and he suggested that these, in fact, were collagen fibers. On closer inspection, there are several flaws in his argument, the most notable being that the integumentary structures of the Liaoning animals are clearly not internal and extend far from the body. Also, the sort of collagen fibers found in the dolphin are associated with aquatic skin types and are also found in sharks and rays, and even as fossils in aquatic reptiles like ichthyosaurs.

This has led someone with the initials JGK to make this lucid comment online: "Unless the author is suggesting (which he did not), that the sort of skin seen in

Along with science, China has a rich erotic tradition of pillow books, sensual sorcery and aphrodisiacs. An assortment of Qing dynasty containers.

cetaceans and sharks evolved, for some spectacularly odd reason, on terrestrial vertebrates like dinosaurs, one is left to wonder just how appropriate his observations really are."

As the evidence piles up, the BAND rhetoric continues to change. The case for feathered dinosaurs is now so persuasive that even the stalwarts of the "birds are not dinosaurs" camp have admitted that the Jehol animals are not fakes and, in fact, have feathers, like modern birds. Yet in the face of evidence that includes anatomical details from all over the skeletons of increasing numbers of specimens, they have not abandoned their belief that birds are not related to dinosaurs. Instead, they have most recently incorporated the Jehol animals into the world of birds, stating it is we who have been mistaken, that these animals with dinosaur-like bodies are not dinosaurs at all; they are, instead, birds. Or, in the words of Ruben, "We now question very strongly whether there were any feathered dinosaurs at all."

Let's review quickly and simply where matters stand.

The relationship between birds and dinosaurs is first suggested in the 19th century. In the 1970s it finds additional support through the work of John Ostrom, which is heavily criticized by the nascent BAND. Additional support is found in Jacques Gauthier's thorough, systematic analysis. This is criticized heavily on the basis of anatomical and developmental subtlety, the temporal paradox, and process arguments about the origin of flight.

New discovery shows even more anatomical similarity between theropod dinosaurs and birds (the presence of a breastbone, wishbones, etc.). BAND dismisses these as convergent evolution, similar structures that evolved independently. *Sinosauropteryx* is discovered. BAND dismisses this as a fake or says the feathers are collagen fibers. Phil Currie points out how this is incongruous with their hollow microstructure.

The first dinosaurs are discovered with feathers of modern aspect. BAND considers them fakes or flightless birds. Unquestionable feathered dromaeosaurs are reported from the Early Cretaceous, significantly closing the time gap between the appearance of the first bird and the appearance of the first dromaeosaur. This diminishes the already rejected idea of a temporal paradox. BAND first considers them fakes, then proclaims that they are actually birds unrelated to other dinosaurs.

This final hypothesis is extremely self-critical. After all of the finger-pointing by BAND adherents about how our data and methods were flawed, many are now ready to accept that dinosaurs like Dave have feathers and wishbones. This is damning criticism of 20 years of their own work pointing out minute differences between dinosaurs like Dave and the skeletons of birds.

My feelings are mirrored by dinosaur artist Kristopher Kripchak in a web posting: "As soon as one theory on why birds cannot be dinosaurs is demolished by a new discovery, the BAND crowd comes up with a new theory that is even less plausible than their previous one. Over the past few years, these folks have adopted more positions than the Kama Sutra."

学者

CHAPTER 11

A Scientist

Information Rendering Much
of This Book Inaccurate

As much as we would like to think that science is cool and calculating and we scientists know exactly where we are going next, it isn't, and we don't. Ideas, the best ones, always come out of nowhere. They come ephemerally at 5 in the morning, when you are exhausted after work, or when the last thing on your mind is science or paleontology.

It is ideas like these that allow us to move forward, to decide where to look for fossils, to understand their significance, and to see the patterns connecting them. Each of these ideas adds to the architecture of knowledge about life on our planet, by placing new discoveries in the context of what is already known. What makes paleontology so special to me is that it shares with other scientific disciplines a methodological tradition going back to the birth of modern Western science 500 years ago, yet retains an element of informed luck and serendipity. As a paleontologist, I am scientist, prospector, opportunist, diplomat, and deal-maker.

In recent years, some of the most important fossils ever unearthed have come from Liaoning. These specimens are crucial to addressing

With thousands of years of recorded history, it is hard not to get a feeling of permanence and great antiquity on any visit to the Middle Kingdom.

As this oil painting by Xu Behong attests, modern Chinese culture, like all cultures, is a peculiar hybrid of the indigenous and the cosmopolitan.

questions that just a few years ago seemed unlikely to be answered. For my part, I have been involved in detailing part of the story of avian origins, sorting out the family tree of theropod dinosaurs, and detailing some of the broad issues about the communities in which these animals lived: how diverse they were and whether differences among localities are due to time or environment. Using the methods of modern systematics and new technology, ideas about the chronology and genealogy of early avian diversification can be tested, as can theories about the evolution of feathers, the origin of flight, behavior, and perhaps the appearance of warm-bloodedness in the progenitors of birds.

One of the most satisfying aspects of my career has been my time working with Chinese scientists, both in their country and mine, studying these remarkable specimens. China is a rapidly evolving country. Gone is the peaceful Beijing of narrow streets shaded by elm trees. Even in the early 1990s, Beijing was still a quiet city, with only the sounds of laughter punctuated by the tinkle of bicycle bells. Much has changed. Impromptu street markets have been replaced by McDonald's and KFC, and their home-grown competitors. Cheese was thought to be toxic to most people, yet now Pizza Hut is one of the most popular restaurants.

Most of the hutongs are gone, replaced by modern apartment blocks, and the bicycle lanes have been narrowed to accommodate Los Angeles-style traffic. The remaining old districts are still there are going through the same sort of gentrification that New Yorkers have seen in brownstone neighborhoods. From a distance, this may seem unfortunate, but Beijing is the capital of an emerging and thriving world power. What capital of an economic superpower has much of the population living packed in 17th century dwellings with no plumbing and only communal toilets?

Beyond the bones, my experiences in China have opened my eyes to a new way of looking at my work, my discipline, and broader issues of politics and society. It has also taught us, Mick and me, that we really do work, sleep, eat, and drink in a global community.

Unlike that of my predecessors, my international work is international. There is a lot more involved than who controls the specimens of feathered dinosaurs. Young colleagues like Xu Xing, Zhou Zhonghe, and Gao Ke-Qin work actively on projects worldwide and present their research at all of the important international congresses. A culture of cooperation is growing and replacing one of colonialism and dogmatic competition.

I ended my previous book, *Discovering Dinosaurs*, published in 1995, with the statement, "We hope that these specimens will provide us with important information that will render much of this book inaccurate." This was in reference to theropod dinosaur fossils collected on Gobi Desert expeditions, when feathered dinosaurs were only imagined. A flurry of activity by dinosaur paleontologists around the world has rendered much of what I wrote in *Discovering Dinosaurs* incorrect. Evidence has been reinterpreted and added to, and theories have been amended, replaced, or shown to be plain wrong.

My times in China studying the feathered dinosaur have been some of the most memorable and enjoyable of my career. Here I examine the slab of the "four fingered theropod" Chong.

In the sixties there was a television show called "Route 66." The two principal characters, Tod and Buzz, cruised the American West in a 1960 Corvette. In the course of discovering America, they encountered outcasts and ordinary people—they were travelers. They discovered much, but they never knew what would happen next, what lay down the road. For us, every trip to China is different. Some things are familiar, while others seem to have changed virtually overnight. We see old friends and colleagues and make new ones. Fossil skeletons from Liaoning continue to be unearthed, and on each visit, we see animals that we never would have imagined existed. Their analysis and incorporation into the larger context of global paleontology is only beginning.

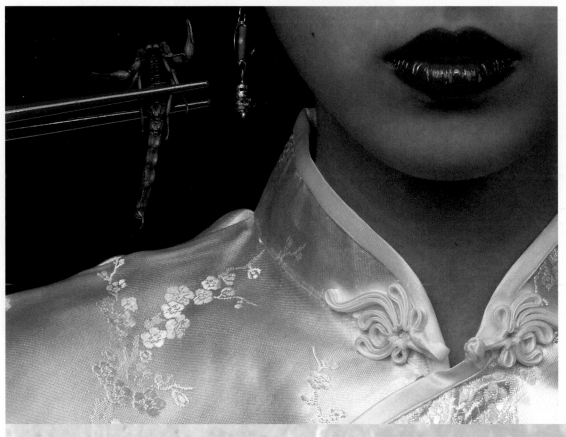

The China that I will always remember is a composite of the beautiful and the unfamiliar.

Tod and Buzz ended each episode with an exhausted Buzz looking over at his compadre:

"Where to now, Tod?"

"I don't know Buzz, just go."

Stay tuned. This trip isn't over, and there's a whole lot out there still to discover.

Notes

INTRODUCTION
Dyed eggshells from the red egg and ginger festival. This celebration is held on the one month birthday of a new child.

CHAPTER 1
Aged paper of an old Chinese schoolbook through a magnifying glass.

CHAPTER 2
The doors on a traditional lacquered chest from Shan Xi.

CHAPTER 3
The deep blue crackled glaze of a Jun kiln bowl. Dating from the Song Dynasty, this sort of glaze has only recently been reproduced.

CHAPTER 4
A detail of the wing of a dragonfly. Dragonflies are often flown on long strings by young children in summer.

CHAPTER 5
The iridescent body of a butterfish, a favorite ingredient in many Chinese dishes.

CHAPTER 6
The teeth of the feathered dromaeosaur *Sinornithosaurus*, perhaps the species to which Dave belongs.

CHAPTER 7
A cast of modern feathers which have been gilded.

CHAPTER 8
A detail of the small feathers or protofeathers which cover the tail of Dave.

CHAPTER 9
Brightly patterned breast feathers of the pheasant *Satyr tragopon*.

CHAPTER 10
Pigeon tracks preserved in concrete on a New York sidewalk. The three toes pointed forward is an indication of their dinosaurian ancestry.

CHAPTER 11
Sihetun emerging from the fog.

ACKNOWLEDGMENTS
Found on the trunk of a several hundred year old cypress tree in the grounds surrounding the Temple of Heaven.

Additional Reading

There is not much outside of the primary technical literature on the feathered dinosaurs of Liaoning and the Jehol Biota. Yet here are a few references that bear on some of the themes in this book.

The Jehol Biota. The Emergence of Feathered Dinosaurs, Beaked Birds and Flowering Plants. Mee-mann Chang et al., Eds. Shanghai Scientific and Technical Publishers, Shanghai, 2004. 210 pp. 350 Yuan. ISBN 7-5323-7318-5. This lavishly illustrated book is the definitive scientific treatment on the Jehol Biota. It is an edited volume compiled by the IVPP.

City of Lingering Splendour : A Frank Account of Old Peking's Exotic Pleasures John Blofeld. Shambhala, 2001. x pp. ISBN 1570626375. A book sensitive and sensual that depicts old Beijing in seductive fashion.

The Search for Modern China. Jonathan D. Spence W. W. Norton & Company; 2nd edition, 1999, 728 pp. ISBN: 0393973514. Although it has been criticized for errors, Spence's book is still the departure gate for anyone wanting to get a short course on today's China.

The Search for Modern China: *A Documentary Collection*. Pei-Kai Cheng, Michael Lestz, Jonathan D. Spence W. W. Norton & Company; 1st ed edition, 1999. 450pp. ISBN: 0393973727. A companion piece to The Search for Modern China that provides translations of relevant historical documents.

The Columbia Guide to Modern Chinese History. R. Keith Schoppa. Columbia University Press, 2000. 320 pp. ISBN: 0231112769. A great reference for information about modern Chinese history.

China Pop: *How Soap Operas, Tabloids and Bestsellers Are Transforming a Culture*. Jianying Zha. New Press 1996. ISBN: 1565842502. Although dated this treatment gives an idea how Chinese culture is evolving through the assimilation of global ideas and creating something distinctly modern Chinese.

Feathered Dinosaurs. Christopher Sloan, Philip J., Dr Currie. National Geographic. 64 ISBN: 0792272196. A children's book, yet with some striking images of one artists portrayal of how some of the feathered dinosaurs may have looked.

A Field Guide to Dinosaurs: *The Essential Handbook for Travelers in the Mesozoic*. Henry Gee, Luis V. Rey. Barron's Educational Series; 1st Us edition. 2003, 144 pages ISBN: 0764155113. A benchmark in that these creatures, rendered at the limit of credulity, are so different than how we would have reconstructed them just a few years ago.

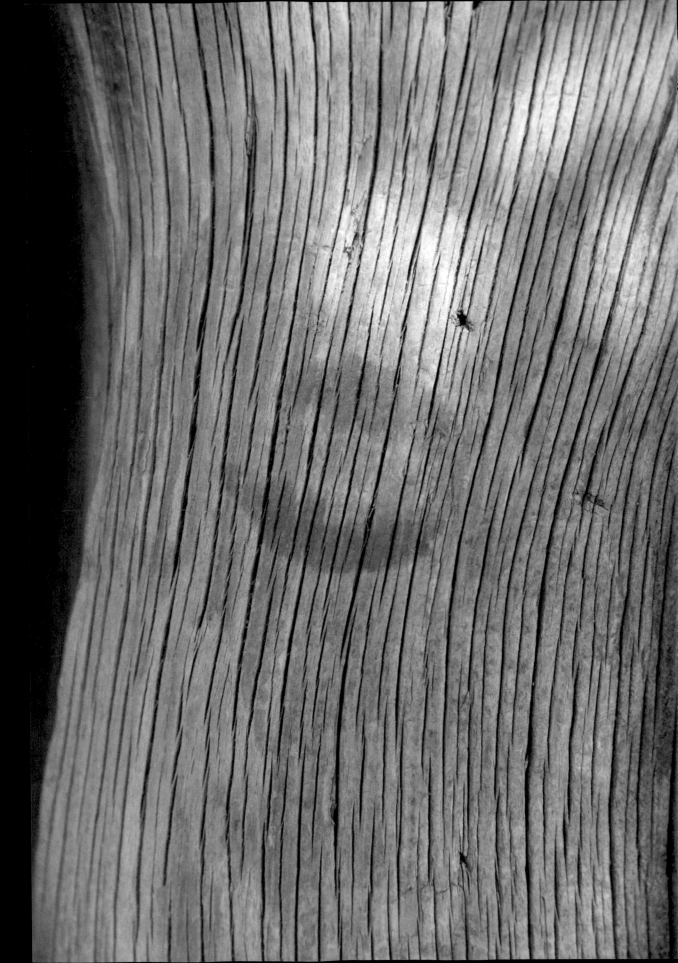

Acknowledgments

M any people both helped us along the way and inspired us to write this book. At home we thank our families Vivian and Inga (Mark) and Marni and Rowan (Mick) for putting up with us. Howard and Ting-Yeh Pan contributed a lot. This project would not have been possible without the support of the American Museum, undoubtedly the finest place on the planet to carry out the kind of work we do. At the museum the members of our division, my colleagues and especially everyone involved in my research group have allowed me to work at a level that made the experiences recounted here possible. Dina Langis, Li Yuexing, Marsha Lee, Iris Zhang, Kala Harinarayanan, and Sunny Hwang were tolerant and beautiful photographic subjects. Carl and Meng thanks for helping us out throughout, and Luis thanks for your sense of humor. Henry and Debbie Galiano, proprietors of Maxilla and Mandible, provided some of the objects for the photographs in this book and Rick Ketcham and Timothy Rowe the *Archaeoraptor* diagram. Special thanks to Billy at Macaleers and the crew at the Dublin House and Red Rocks, and in Beijing at the Buddha Bar 2, the Xiyuan Hotel, and the ghosts of the old city that still inhabit it.

In China our thanks go out to too many to name individually, however it would not have been possible without Gao Ke-Qin and Ji Qiang. The next Baijiu is on us. Several at the IVPP have contributed greatly to the research and our understanding in China especially Mee-Man Zhang, Xu Xing and Zhou Zhonghe. Dong Zhiming, for introducing me to snake restaurants on my early visits and giving me my first glimpses of Chinese dinosaurs. And we can't forget Ren Dong at Capital Normal University, and Ji Shu-An. Bao, Makai, Lei Yu, Helen you are great friends who have taught us a lot about modern China. Gao Hui,

Tiger's Mom, Qing Qing, Uncle An, Xiao Hu, and Mr. Number 5 you will never be forgotten.

Our editor Stephen Morrow gave some of the writing focus, Peter N. Névraumont for producing the book, Cathleen Elliott for the book's design, and Jean Christensen for copy editing.

Lastly we thank the people of China. It is our hope that our countries have a great future in navigating the road ahead.

Index